# THE NORTH AMERICAN MODEL OF WILDLIFE CONSERVATION

*Wildlife Management and Conservation*

Paul R. Krausman, Series Editor

# THE NORTH AMERICAN MODEL OF WILDLIFE CONSERVATION

EDITED BY

Shane P. Mahoney and Valerius Geist

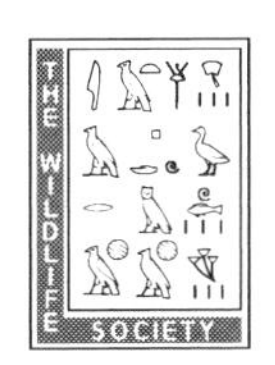


Published in Association with *THE WILDLIFE SOCIETY*

JOHNS HOPKINS UNIVERSITY PRESS | BALTIMORE

Printed in the United States of America on acid-free paper
9 8 7 6 5 4 3 2 1

Johns Hopkins University Press
2715 North Charles Street
Baltimore, Maryland 21218-4363
www.press.jhu.edu

Library of Congress Cataloging-in-Publication Data
Names: Mahoney, Shane P., 1956– editor.
Title: The North American model of wildlife conservation / edited by Shane P. Mahoney and Valerius Geist.
Description: Baltimore : Johns Hopkins University Press, 2019. | Series: Wildlife management and conservation | Includes bibliographical references and index.
Identifiers: LCCN 2019001478 | ISBN 9781421432809 (hardcover : alk. paper) | ISBN 1421432803 (hardcover : alk. paper) | ISBN 9781421432816 (electronic) | ISBN 1421432811 (electronic)
Subjects: LCSH: Wildlife conservation—North America.
Classification: LCC QL84 .N67 2019 | DDC 333.95/40973—dc23
LC record available at https://lccn.loc.gov/2019001478

A catalog record for this book is available from the British Library.

*Special discounts are available for bulk purchases of this book. For more information, please contact Special Sales at 410-516-6936 or specialsales@press.jhu.edu.*

Johns Hopkins University Press uses environmentally friendly book materials, including recycled text paper that is composed of at least 30 percent post-consumer waste, whenever possible.

# Contents

# Acknowledgments

An edited volume always entails the inevitable turns of circumstance that prevail in the personal and professional lives of its contributors. Indeed, there is always a certain sense of adventure in seeking the input of colleagues whose own careers are often overflowing and who must be pursued for the immediate task of writing to volume deadlines, across the turbulent social and intellectual environments they inhabit. However, as co-editors for the volume in hand, we can only express our gratitude for the thoughtful perseverance and quiet patience of a wonderful assembly of colleagues and take every responsibility for any delays or detours encountered along the road to this book's publication.

We would also wish to thank the various editorial staff at Johns Hopkins University Press, including especially editors Vincent Burke and Tiffany Gasbarrini, for their many assistances, promptings, and encouragements throughout this process. Likewise, we thank Amanda Hull of Conservation Visions, who ably assisted in various ways to keep the flow of information going when the co-editors were themselves overwhelmed with work and travel schedules. Our appreciation is also and especially extended to Dr. Paul Krausman, editor for the Wildlife Society–Johns Hopkins University Press Wildlife Book Series for his constant readiness to assist with every request made of him and for his thoughtful advice and encouragement, always.

In reviewing the many contributions to this volume, we would be remiss if we did not also acknowledge the many conservation leaders, some well known but many others not, who over the last one hundred years laid the foundation for conservation in Canada and the United States. In this first book to examine that incredible and lasting contribution to social and environmental progress, we remain humbled by the efforts they made. This book is the product of our labors today, certainly; but, even more so, it is the legacy of incredible labors long ago. Without such efforts, no North American Model would exist and no book would be dedicated to its cause.

# About the Contributors

*Leonard A. Brennan* is a Professor and C. C. Charlie Winn Endowed Chair for Quail Research at the Caesar Kleberg Wildlife Research Institute at Texas A&M University–Kingsville.

*Rosie Cooney* is an independent consultant and Senior Affiliate with the Fenner School of Environment and Society at The Australian National University.

*James L. Cummins* is a certified wildlife biologist and Chief Executive Officer of Wildlife Mississippi.

*Kathryn Frens* is a Ph.D. student in the Environmental Science and Policy Program at Michigan State University and a National Oceanic and Atmospheric Association Sea Grant Knauss Fellow.

*Valerius Geist* is Professor Emeritus at the University of Calgary. He first articulated the North American Model of Wildlife Conservation in 1995.

*James R. Heffelfinger* is a Full Research Scientist and Wildlife Science Coordinator at the Arizona Game and Fish Department.

*David G. Hewitt* is a Research Scientist and Executive Director of the Caesar Kleberg Wildlife Research Institute at Texas A&M University–Kingsville.

*Paul R. Krausman*, Professor Emeritus at the University of Arizona, is Editor in Chief of the *Journal of Wildlife Management*.

*Shane P. Mahoney* is President of Conservation Visions and Deputy-Chair of the International Union for Conservation of Nature's Sustainable Use and Livelihoods Specialist Group.

*John F. Organ* is a former Chief of the United States Geological Survey, Cooperative Fish and Wildlife Research Units.

*James Peek* is Professor Emeritus at the University of Idaho's Department of Fish and Wildlife Science.

*William Porter*, a Boone and Crockett Professor of Wildlife Conservation, leads the Boone and Crockett Quantitative Wildlife Center at Michigan State University.

*John Sandlos* is a Professor at Memorial University of Newfoundland's Department of History.

*James A. Schaefer* is a Professor of Biology at Trent University.

# 1 The North American Model of Wildlife Conservation

## *Setting the Stage for Evaluation*

SHANE P. MAHONEY,
VALERIUS GEIST, AND
PAUL R. KRAUSMAN

*This chapter introduces the North American Model of Wildlife Conservation, including its basic principles and their significance within the framework of conservation achievement. Support mechanisms such as public engagement and funding are also referenced. The authors reflect on the Model's strengths and weaknesses and consider its continued relevance and efficacy in terms of historical, political, cultural, and social realities.*

### Introduction

The conservation of wildlife is surely one of humanity's most complex challenges, and overwhelming evidence indicates that the loss of wild species is escalating rapidly at a global scale.[1] The reasons for this are numerous and varied, but they ultimately derive from conflict between the finite capacity of natural systems to support diversity and the ever-increasing demands for resources and space by one species, *Homo sapiens*. The search for solutions to this dilemma preoccupies large numbers of governmental and nongovernmental bodies and an army of practitioners worldwide. Yet those solutions remain elusive, particularly at large temporal and geographic scales. Nevertheless, by exploring existing approaches, we can illuminate the answers we seek and the mistakes we wish to avoid.

This book is about one such approach and is, in fact, the first book to focus on the North American Model of Wildlife Conservation. But what is the Model exactly? The Model is not an article of constitution. It is an evolved and shared system of conservation laws, principles, institutions, and policies that has enabled the successes of Canada and the United States in the recovery, management, and protection of wildlife and brought them global recognition. Like all such approaches, North America's arose within and *because of* particular historical, cultural, social, and political realities. In hindsight, it is always possible to view such outcomes as inevitable consequences, but this tendency masks some of the most remarkable elements of North America's conservation achievements. It also blinds us to some of its shortcomings. This book is intended to explore both and comes at a time when numerous principles of the Model are being challenged, when vacancies in the Model's described architecture are being discussed, and when social institutions that support or oppose it are rapidly changing.

### A Revolutionary Approach

Launched in the late nineteenth and early twentieth centuries, the partnered Canada–United States approach to wildlife conservation was revolutionary for its time. Rather than treating wildlife as the purview of elites, to be used for their benefit and pleasure, as it had been in many parts of Europe historically, the new approach considered wildlife public property, accessible for legal purpose—to harvest, view, or otherwise enjoy—by any citizen in good standing. It thus freed wildlife from arbitrary, private control and eventually placed its fate in the hands of professionals educated in institutions of higher learning, as well as in the hands of government officials elected to serve the public and manage this resource on their behalf. It provided solid conservation funding through innovative laws, passed by all levels of government. It accepted international cooperation and established treaties for managing migratory birds and other wildlife species. It largely removed market policies harmful to wildlife, controlled trade in wildlife through enforceable laws, and led to the establishment of numerous organizations specifically dedicated to and promoting the welfare of a majority of harvested wildlife (game) species.

These were remarkable innovations and achievements, many introduced over a century ago and guided by a gradually evolving system of principles that saw maturity by the 1930s but were first fully articulated by Valerius Geist in 1995.[2] These principles, which form the conceptual and philosophical basis for the North American Model, have been variously described in subsequent peer-reviewed publications and have been popularly referenced as the Seven Sisters of Conservation.[3] While the precise wording varies somewhat, the list, intentions, and relevance of these principles are consistently represented. The following descriptions are from Mahoney and Jackson "Enshrining Hunting," where a detailed discussion of the importance and implications of each principle is presented.[4]

1. Maintaining wildlife as a public trust resource, entrusted to the state to manage.
2. Prohibiting deleterious commerce in dead wildlife products.
3. Regulating and defining appropriate wildlife use by law.
4. Ensuring wildlife can only be killed for legitimate purpose.
5. Recognizing and managing wildlife as an international resource.
6. Utilizing and safeguarding science as the appropriate basis for wildlife policy.
7. Protecting the democratic allocation of citizen opportunity to harvest wildlife.

The collective impact of these principles was to rescue and restore many imperiled species to abundance in North America and to establish and maintain a unique approach to wildlife conservation that is one of the great achievements of Canadian and American societies. In addition to its biological successes, the approach fostered a massive associated economy that has brought livelihood benefits to millions of people across more than three generations and that continues to incentivize investment and innovation. At the same time, and as a crucial component of its structure, this conservation system has sustainably provided access to massive wild meat harvests and spawned a vast array of organizations dedicated to wildlife and its sustainable use. Implemented and applied through state/provincial and federal government agencies, this long-standing, professional approach to wildlife management was born at a time when wildlife was being depleted on a massive scale in North America and today manages a wildlife abundance that would have been very difficult to envisage at that time.

Thus, the North American Model does, in many ways, provide a hopeful vision for today's and tomorrow's conservation challenges. As numerous chapters in this book reveal, it is possible to reverse even continental declines in natural diversity and abundance while maintaining multiple uses and access to wildlife resources. Indeed, the North American Model has been incredibly successful in returning a variety of highly valued wildlife species from precarious scarcity to high abundance, reintegrating and safeguarding them in American and Canadian environments. Of equal importance is that broad principles of conservation can be applied across cultural and political divisions. Thus Canada, a British colony at the time the Model was coalescing in the early twentieth century, joined the United States in an unproven, speculative venture in wildlife conservation a century ago, which grew into the Model as we know it today. Its application exemplifies the fact that human diversity and social complexity are not insurmountable barriers to conservation success and that a concern for wildlife's welfare can unite disparate groups in fruitful cooperation. These are insights of great relevance, certainly. But how was all this possible? What was the historical and social context for the Model's emergence?

## Wildlife Matters

Beyond the brilliance and determination of conservation leaders in both countries during the Model's formative stages, what was it that enabled and sustained this new conservation order? Fundamentally, wildlife mattered to the American and Canadian publics, and securing its welfare was based on a broad consensus and supportive policies. This is a crucial point and must not be lost in the detailed explorations of the Model's institutions, principles, and policies in the chapters that follow. Essentially, wildlife was and still is protected by the good will and laws of the American and Canadian citizenry. It is also protected through wide public acceptance of North America's wildlife management approach, a major focus of which is to allocate wildlife for personal consumption and to do so by lawful right of individual access, rather than through the market place. Under this approach, citizens are free, should they choose, to access and enjoy food of very high quality, an issue of growing relevance in modern society. The importance of such access to the emerging and increasing concerns for food security and human health cannot be dismissed and are surely important to the Model's future relevance. Furthermore, the specific conservation benefits that have derived from this aspect of the Model, which safeguards the right of citizens to access wild protein, should not be underestimated. As indicated below, this principle has encouraged a community of users who have been crucial to funding and otherwise supporting conservation in North America. More recently, this principle and especially its application within the Model are coming under increased scrutiny and some legitimate criticism.

## Incentivized Conservation

This focus on the sustainable use of wildlife for consumption has given rise to supporting industries managed by the private sector (e.g., hunting clubs, guides, and outfitters, clothing, ammunition, and gun manufacturers) while generating substantial wealth and employment across diverse sectors of (especially) local economies. Such economic outcomes further incentivize support for sustainable use conservation policies and help create influential constituencies focused on wildlife's future. Furthermore, those who choose to harvest wildlife and fish under the lawful prescriptions of state and federal agencies are themselves incentivized to support wildlife conservation programs. This linkage between conservation success and benefits to people from the use of wildlife was forged very early as a foundation of the Model and provides an important international context for this book. Indeed, integrating local economies and livelihoods in conservation prescriptions is considered a priority by many of today's leading conservation institutions and is recognized globally as a vital component for successful conservation planning.[5]

## Examining the Model

Despite such achievements, however, the Model, both as a historical narrative for understanding the origins and gradual development of North American conservation and as a prescription for future conservation success, requires thoughtful inspection. We should recall that while some wildlife species fared well under the Model's prescriptions, many others did not. In fact, many North American species became extinct in the twentieth century, as the recovery of "game" or harvested wildlife proceeded in spectacular fashion. True, most of those lost were less visible invertebrates and species found in aquatic habitats that might have escaped notice,[6] but there were also abundant terrestrial vertebrate species such as Bachman's Warbler (*Vermivora bachmanii*)[7] and the Eskimo Curlew (*Numenius borealis*)[8] that disappeared during that time. Where in the Model does their fate reside? Where, in the many supportive publications about the Model, is their fate considered?

This disruptive pattern, where some species fared well but others did not, helps reveal both strengths and weaknesses of the Model, and both are important considerations for this book. So, too, is reflecting on our very characterization of conservation history and whether the story told is accurate and complete. Have we reinvented or distorted history to reflect a particular view of conservation? Examining such potential biases is critical for assessing the Model's relevancy and for ensuring that historical evidence is used effectively and impartially to improve future conservation and management efforts, in North America and elsewhere. The historical context of the Model presented later in this book does provide some insight into why the North American system developed as it did and an explanation for policies that govern the use and management of wildlife in North America today.[9] But this context also helps us realize that conservation is never complete. It is an ongoing problem requiring ongoing efforts that play out in a dynamic social reality. This challenge may also be said of our reflections on conservation's rise—there is always more to be discovered, more to add. To approach the truth about conservation's past, about the reasons for its inception and the rationale for its current form, requires vigilance and must be based on ongoing reassessments and the application of science and scholarship. Critical assessments are needed if we are to understand conservation and to help this human endeavor succeed.

And such scrutiny must apply to the North American Model itself. The Model cannot become an orthodoxy, nor should questioning it be viewed as a violation. Indeed, we may well ask whether the fate of Bachman's Warbler, the Eskimo Curlew, and a host of other species can, in any way, be interpreted as a consequence of the Model's focus on a restricted guild of game or hunted species. Thanks to recreational hunters and anglers and their contributions to conservation, harvested species have generally made remarkable recoveries in Canada and the United States, and their populations are mostly stable or increasing in size today.[10] But these were the species focused on during much of the twentieth century, and they still are prioritized today. Is it unreasonable to suggest that the disproportionate attention given to such species by state and provincial agencies may have affected capacities to launch appropriate management actions that could have prevented the extinctions of numerous other species?[11] Maybe. Maybe not. Certainly, what we can say is that, traditionally, programs for nonharvested wildlife species are not as well funded in North America. Moreover, nonhunted species have received less conservation and management attention, overall, and are often less protected from a legislative point of view.[12]

For example, despite it being one of the principles of the Model that markets for wildlife should be eliminated, the commercial fur trade was never impeded in Canada or the United States, even though some species, such as beaver, were decimated by the activity. Furthermore, commercial trade in reptiles and amphibians, primarily for the pet industry, persists in both countries.[13] It would be inconceivable to allow elk, caribou, and other hunted species in today's North America to be subject to the marketplace or right of ownership. Thus, while wildlife management policies should, ideally, apply to all wildlife taxa, this has not been the case in North America and Canada, where stakeholders and advocacy bases have shaped legislation and funding.[14] It is not surprising, therefore, that sustainable wildlife management in North America can appear to sectors of the wider public as self-serving, since efforts preferentially target exploited species *because they are exploited*. This, too, is a critical perspective for exploring the writings presented in this book and for assessing the Model's ongoing relevance to wildlife conservation. Public support is critical.

Along with prioritizing hunted species, the Model has necessarily prioritized the views of consumptive users. Those groups not traditionally engaged in hunting and angling (such as women, some cultural minorities, and appreciable percentages of urban residents) have often been excluded when wildlife policies were being developed, though this is changing.[15] Perhaps most regrettably, the Model has never emphasized nor acknowledged the systems of wildlife use and habitat management that diverse Native peoples had established long before European colonization. The North American Model we recognize today is, of course, a European immigrant construction that was both required and made possible by the destruction of the continent's pre-Columbian wildlife abundance and, later, its extraordinary diversity of human cultures. The ecological views of these peoples and their incomparable knowledge of wildlife were never incorporated within the Model, a reality fraught with consequences. Today, tension often exists between the continent's indigenous or First Nations communities and other users of wildlife. From the latter's perspective, indigenous rights to hunt and fish can be viewed as disproportionate or preferential, though indigenous peoples see such rights as only natural. This tension cannot be remedied without reference to the historical realities that gave rise to it. In the meantime, it is clear that in spite of the Model's key principle of democratic access to hunting opportunities, the reality is that legal access to wildlife for harvest and consumption on the part of indigenous and settler communities is uneven, and sometimes dramatically so.

In reviewing the Model, it is thus important not only to consider its successes, but also to expose its limitations and weaknesses, both from a historical perspective and in view of future endeavors in conservation. This book emerges at a time when even the most fundamental attributes of the Model are under scrutiny and debate. For example, critics have suggested that there is actually insufficient use of science in wildlife management decision-making, despite the importance placed on such knowledge as a core foundation of the Model.[16] Proponents of the Model, meanwhile, contend that science is regularly and effectively applied in wildlife management efforts across North America.[17] Viewed in a wider context, of course, conservation policy decisions should not be based on science alone, but must also consider conservation ethics and other human values, and it is for not embracing these wider considerations that the Model has been most often and effectively criticized.[18]

Since the inception of its conservation movement over one hundred years ago, North America has witnessed vast cultural and economic change. In particular, increased urbanization and decreased personal engagement with animal death have disconnected many people from the origins of their food.[19] This is leading to substantial change in society's general attitudes toward animals, our relationships with them, and acceptance of using them for human purposes, including for food. A rising empathy for animals and an increasing emphasis on vegetarian lifestyles are taking place and can lead to a decreased awareness of how many people worldwide and in North America still rely on animals, wild and domestic, as significant components of their diets and livelihoods, and a general lack of appreciation for the importance of sustainable wildlife management efforts overall.[20] This often leads people to question whether any management of wildlife is required at all. Moreover, not only has there been a reduction in participation in activities long supportive of the Model, such as recreational hunting and angling, but social acceptance of these practices appears to be waning, certainly internationally. No successful exploration of the Model can ignore such realities or neglect to take them into account going forward.

## Animal Welfare and Animal Rights

Indeed, in North America and worldwide, there is no doubt that animal welfare and animal rights movements have gained public and political support over recent decades.[21] These movements are variously opposed to animal use for food, although consumption of meat worldwide is presently increasing, and for other purposes, such as for fur or entertainment. These changing social values will likely grow in importance, escalating the influence of these movements over time, freighting predictable and increasing tensions over humanity's relationship with wildlife and increasing society's tendency to reduce human use of wildlife for food and other purposes. Such social perspectives will predictably lead to increased debate over North America's conservation approach, emphasizing as it does consumptive use of wildlife and influence how relevant the Model itself will remain. Indeed, this process is already established and some of its predictable outcomes already manifest, as growing opposition to hunting of carnivores and other iconic wildlife species indicate.

## Declines in Recreational Hunting Participation

Participation in recreational hunting in Canada and the United States has been declining at a significant rate for decades, despite many ongoing efforts to encourage hunter retention and recruitment.[22] There has been a net loss of approximately 2.2 million hunters in the United States alone since 2011.[23] Between 1991 and 2011, the proportion of US hunters aged sixteen to forty-four decreased from 71 percent to 45 percent, while the proportion of older hunters increased substantially.[24] Approximately one-third of all US hunters today range in age from 54 to 72 years old.[25] Trends are similar in Canada. Approximately 8 percent of Canadians participated in hunting in 2012, down from 16 percent in 1991.[26] It is very likely these trends will continue.

Since funding for North American conservation and wildlife management programs is so closely linked to participation in hunting, especially in the United States, there has been a decrease in funding available to state agencies as a result of the decline in hunter participation.[27] This trend, left unchecked, has the potential to substantially change the current system of wildlife management and, therefore, the Model itself. Successful recruitment and retention of hunters are essential for the hunting tradition to continue and for hunting to remain a viable tool, incentive, and agency for sustainable wildlife management.[28] Thus they are critical to the Model, at least as it is applied and structured today. While there have been numerous attempts to develop alternative and/or supplementary funding mechanisms for conservation in North America, and the United States especially, none has yet been established at the national or continental scale. If and when this is achieved, the influence of hunters in wildlife conservation policy will predictably decrease.

## Conclusion

This, then, is the context in which this book emerges. Current social and economic realities are vastly different from those that prevailed when the Model arose over a century ago. From our attitudes toward animals, to our heightened awareness of environmental concerns such as climate change and biodiversity loss, to the very make-up of Canadian and American societies with respect to cultural diversity, the growth of cities and loss of rurality, and the dynamics of gender relationships, we live in a different world from our ancestors of generations ago. While the chapters to follow shed light on an extraordinary undertaking, one far ahead of its time in many ways, there can be little doubt that the North American Model needs to address challenges that include its restricted species guild emphasis (namely, game and fur-bearing species) and its exclusion of and sometimes antagonism toward wider constituencies (which were part of its founding) such as the wilderness and protected areas movements. It must also attempt to engage more directly and thoughtfully with immigrants, minorities, urban communities, and indigenous peoples. Conservation approaches, to be progressive, must continuously innovate. They must also coordinate with science and scholarship and law and law enforcement, as well as take into account wildlife and nature economics.

As the discussions in the ensuing chapters emphasize, the Model was built, especially in its early years, on an appeal to the citizenry, which led to the formation of prideful constituencies who defined themselves as conservation advocates. For the Model to work and for wildlife to thrive, it must be seen as a both a utilitarian and an emotional component of social discourse so as to foster sound conservation approaches. The public must be continuously engaged, but this public is, of course, changing, and therein lies the Model's greatest challenge: Can it adapt fast enough and yet secure the basic principles and effective mechanisms required to keep public support and wildlife abundance at the same time? Over the course of one hundred years, no other group besides hunters and anglers has stepped up to the financial commitment challenge to make conservation work in North America. Now, though, the question is whether existing energies for conservation are sufficient to this purpose. We must ask ourselves *for whom* is wildlife managed and *for what purpose*.

These are not new questions, of course, though they are now reaching a critical intensity for conservation policy makers. Surely it is impossible—and dangerous—to ignore that we are now at a crossroads, a tipping point where significant change is inevitable. What has not changed, perhaps, is the fundamental question of who will care sufficiently to pay for wildlife conservation in the future.

This book cannot address every issue in North American conservation history or cover every emerging challenge. Its purpose is to present readers with the widest and most detailed coverage to date of the North American Model. The book's broader purpose, however, is to elicit thoughtful debate, not only about the Model, its guiding philosophy, history, successes, omissions, and challenges, but also about wildlife's future in Canada and the United States. We hope that the chapters presented here will serve as incentive for this wider discussion and for conceiving the road ahead for wildlife conservation on this continent.

*NOTES*

1. M. Grooten and R. E. A. Almond, eds., *Living Planet Report—2018: Aiming Higher* (Gland, Switzerland: WWF, 2018).
2. V. Geist, "North American Policies of Wildlife Management," in *Wildlife Conservation Policy*, ed. V. Geist and I. McTaggert-Cowan (Calgary: Detselig, 1995), 77–129.
3. V. Geist, S. P. Mahoney, and J. F. Organ, "Why Hunting Has Defined the North American Model of Wildlife Conservation," *Transactions of the North American Wildlife and Natural Resources Conference* 66 (2001): 175–185; J. F. Organ, S. P. Mahoney, and V. Geist, "Born in the Hands of Hunters: The North American Model of Wildlife Conservation," *Wildlife Professional* 43, no. 3 (2010): 22–27; S. P. Mahoney and J. J. Jackson, III, "Enshrining Hunting as a Foundation for Conservation—The North American Model," *International Journal of Environmental Studies* 72, no. 5 (2015): 869–878; J. F. Organ, V. Geist, S. P. Mahoney, S. Williams, P. R. Krausman, G. R. Batcheller, T. A. Decker, R. Carmichael, P. Nanjappa, R. Regan, R. A. Medellin, R. Cantu, R. E. McCabe, S. Craven, G. M. Vecellio, and D. J. Decker, "The North American Model of Wildlife Conservation," *Wildlife Society Technical Review* 12, no. 4 (2012): viii–ix; S. P. Mahoney, "Recreational Hunting and Sustainable Wildlife Use in North America," in *Recreational Hunting, Conservation and Rural Livelihoods: Science and Practice*, ed. B. Dickson, J. Hutton, and W. M. Adams (Hoboken, NJ: Wiley-Blackwell, 2009), 266–281.
4. Refer to Mahoney and Jackson for a detailed discussion of the importance and implications of the seven principles of the North American Model of Wildlife Conservation. Mahoney and Jackson, "Enshrining Hunting."
5. N. Salafsky and E. Wollenberg, "Linking Livelihoods and Conservation: A Conceptual Framework and Scale for Assessing the Integration of Human Needs and Biodiversity," *World Development* 28, no. 8 (2000): 1421–1438.
6. B. Collen, M. Böhm, R. Kemp, and J. E. M. Baillie, *Spineless: Status and Trends of the World's Invertebrates* (London: Zoological Society of London, 2012).

7. H. M. Stevenson, "The Recent History of Bachman's Warbler," *Wilson Bulletin* 84, no. 3 (1972): 344–347; P. B. Hamel, "Bachman's Warbler (*Vermivora bachmanii*), Version 2.0," The Cornell Lab of Ornithology, Birds of North America, ed. A. F. Poole, August 19, 2011, https://birdsna.org/Species-Account/bna/species/bacwar/introduction.
8. C. A. Fannes and S. E. Senner, "Status and Conservation of the Eskimo Curlew," *International Council for Bird Preservation* 45, no. 2 (1991): 237–239.
9. Mahoney and Jackson, "Enshrining Hunting."
10. C. H. Flather, G. D. Hayward, S. R. Beissinger, and P. A. Stephens, "Minimum Viable Populations: Is There a 'Magic Number' for Conservation Practitioners?" *Trends in Ecology and Evolution* 26, no. 6 (2011): 307–316.
11. M. N. Peterson and M. P. Nelson, "Why the North American Model of Wildlife Conservation Is Problematic for Modern Wildlife Management," *Human Dimensions of Wildlife* 22, no. 1 (2016): 43–54.
12. M. J. Peterson, M. N. Peterson, and T. R. Peterson, "What Makes Wildlife Wild? How Identity May Shape the Public Trust versus Wildlife Privatization Debate," *Wildlife Society Bulletin* 40, no. 3 (2016): 428–435.
13. Organ et el., "The North American Model of Wildlife Conservation," 25.
14. Peterson and Nelson, "Why the North American Model of Wildlife Conservation Is Problematic."
15. A. M. Feldpausch-Parker, I. D. Parker, and E. S. Vidon, "Privileging Consumptive Use: A Critique of Ideology, Power, and Discourse in the North American Model of Wildlife Conservation," *Conservation and Society* 15, no. 1 (2017): 33–40; Peterson and Nelson, "Why the North American Model of Wildlife Conservation Is Problematic."
16. K. A. Artelle, J. D. Reynolds, A. Treves, J. C. Walsh, P. C. Paquet, and C. T. Darimont, "Hallmarks Missing from North American Wildlife Management," *Science Advances* 4, no. 3 (2018), https:// DOI: 10.1126/sciadv.aao0167.
17. Organ et al., "The North American Model of Wildlife Conservation"; D. J. Decker, A. B. Forstchen, W. F. Siemer, C. Smith, R. K. Frohlich, M. Schiavone, P. Lederle, and E. Pomeranz, "Moving the Paradigm from Stakeholders to Beneficiaries in Wildlife Management," *Journal of Wildlife Management* (2019), https://doi.org/10.1002/jwmg.21625; S. J. Riley, J. K. Ford, H. A. Triezenberg, and P. E. Lederle, "Stakeholder Trust in a State Wildlife Agency," *Journal of Wildlife Management* 82, no. 7 (2018): 1528–1535, https://doi.org/10.1002/jwmg.21501.
18. Decker et al., "Moving the Paradigm."
19. M. L. McKinney, "Urbanization, Biodiversity, and Conservation: The Impacts of Urbanization on Native Species Are Poorly Studied, But Educating a Highly Urbanized Human Population about These Impacts Can Greatly Improve Species Conservation in All Ecosystems," *Bioscience* 52, no. 10 (2002): 883–890.
20. S. G. Clark and M. B. Rutherford, *Large Carnivore Conservation: Integrating Science and Policy in the North American West* (Chicago: University of Chicago Press, 2014); S. P. Mahoney, ed., *Monograph: Conservation and Hunting in North America II* (London: Routledge, 2015).
21. J. L. Anderson, "Protection for the Powerless: Political Economy History Lessons for the Animal Welfare Movement," *Stanford Journal of Animal Law and Policy* 4, no. 1, Drake University Law School Research Paper No. 11-09 (2011): 1, https://ssrn.com/abstract=1946337.
22. N. Rott, "Decline in Hunters Threatens How U.S. Pays for Conservation," National Public Radio, March 20, 2018, https://www.npr.org/2018/03/20/593001800/decline-in-hunters-threatens-how-u-s-pays-for-conservation.
23. US Department of the Interior, US Fish and Wildlife Service, and US Department of Commerce, US Census Bureau, "2016 National Survey of Fishing, Hunting, and Wildlife-Associated Recreation," *United States Census Bureau Library*, October 24, 2018, https://www.census.gov/library/publications/2018/demo/fhw-16-nat.html.
24. N. Krebs, "Why We Suck at Recruiting New Hunters, Why It Matters, and How You Can Fix It," *Outdoor Life*, January 18, 2018, https://www.outdoorlife.com/why-we-are-losing-hunters-and-how-to-fix-it.
25. Krebs, "Why We Suck at Recruiting New Hunters."
26. H. Albrecht, "Licensed Hunting and Trapping in Canada. Report of the Standing Committee on Environment and Sustainable Development," *Government of Canada Publications*, June 1, 2015, http://publications.gc.ca/collections/collection_2015/parl/xc50-1/XC50-1-1-412-10-eng.pdf; A. H. Macpherson, "Hunting," *The Canadian Encyclopedia*, March 4, 2015, https://www.thecanadianencyclopedia.ca/en/article/hunting.
27. J. R. Heffelfinger, V. Geist, and W. Wishart, "The Role of Hunting in North American Wildlife Conservation," *International Journal of Environmental Studies* 70, no. 3 (2013): 399–413.
28. Krebs, "Why We Suck at Recruiting New Hunters."

# 2

Valerius Geist and
Shane P. Mahoney

# North American Ecological History as the Foundation of the Model

*The North American Model of Wildlife Conservation represents a dynamic framework of policies and laws that have evolved over time. Consequently, some knowledge of North America's wildlife and human history is necessary for understanding the origins and architecture of the Model itself. Not to be overlooked here are the oral traditions of Native Americans, such as those represented by Larry Littlebird from Pueblo culture in New Mexico.*[1] *Discussions of the evolving nature of wildlife conservation efforts in Canada and the United States also need to take into account how the concepts of naturalness and protectionism may be at odds with some of the basic Model principles that have been so important for past conservation successes. In this chapter we outline North America's ecological history in broad terms, providing a basis for understanding the more detailed explorations of relevant issues in the chapters to follow.*

## An Introduction to North American Wildlife Conservation

Wildlife in North America is subject to very diverse policies. Some have to do with managing wildlife for consumption and using public lands for livestock grazing;[2] others are related to the highly complex and sometimes contradictory protection and management policies and processes of the large national park and protected areas systems;[3] and still others depend on policies and priorities of military reserves, as well as the desires of owners of extensive private lands.[4] These policies are under scrutiny by a more or less informed, but opinionated and politically active segment of the public, much of it organized into societies that, purportedly, want to protect "nature." The North American Model of Wildlife Conservation,[5] for all its virtues and vices, cannot be discussed in isolation from such powerful claims on America's wildlife and purported best conservation practices.[6] An all too common denominator of these claims, as well as of those concerning rewilding,[7] is an implicit or explicit reference to some idea of what is "natural." The term "rewilding" is a widely used term today denoting efforts to restore wildness conditions to landscapes, or generating the wolf form from a domestic species, such as rewilding cattle to the parent form, the extinct urus. Consequently, in exploring the history and relevance of the North American Model it behooves us to examine what is natural, and whether North America's objectives for wildlife reflect realistic interpretations of the continent's past and hopes for its future.

## A Brief History of North American Wildlife

The history of North American wildlife goes back to the late Pleistocene, when over sixty genera of large-bodied mammals, giant birds, and reptiles inhabited the continent. Given that many of these were tree crunchers, including mammoths, two species of mastodon, and four species of giant ground sloths and others were shrub-eaters, including camels, two species of llamas, four or five species of tapirs, a shrub ox, forest musk oxen, stag moose, large peccaries, white-tailed and black-tailed deer, and several species of pronghorns, the landscape must have been closely cropped and rather open. Only under such conditions could the many species that had cursorial, that is, high-speed running adaptations, evolve. Foremost among these was the pronghorn, as well as the giant bison, which was geared for speed rather than power as is the current bison. Other species of note were horses and onagers, long-legged carnivores, such as the huge short-faced bear, the scimitar cat, the American cheetah, and the Colombian mammoths, which had exceptionally long legs, for only leg length allows for speed in elephants. There were giant vultures, now extinct, and big armored beasts, such as giant tortoises and enormous glyptodonts, some with spiked tail-clubs mimicking *Ankylosaurus* from the age of dinosaurs. And there was a plethora of large, specialized, and very assertive carnivores.

However, the bulk of large-bodied species were plant eaters that devoured everything from trees to lichens, opening up the landscape, while also ensuring the evolution of efficient defenses by plants against being eaten, such as thorns, tough-fibered branches resistant to breakage, as in the many "ironwooded trees," and an unending array of poisonous properties. It is little wonder, therefore, that the most toxic tree in the world, the Machineel tree (*Hippomane mancinella*), grows in southern North America (currently it is endangered in Florida); so does the poisonous Sandbox tree (*Hura crepitans*), whose sap is half a million times more toxic than potassium cyanide.[8] Also think of poison ivy, poison oak, poison sumac, wolf bane, deadly nightshade, devil's snare, and so many others. Moreover, the continent's ecology was not static, but moved with the great climatic changes during the glacial cycles and included widespread extinctions long before humans set foot in North America.[9]

Early humans entered eastern Siberia and thus stood at the gate of Alaska for over 800,000 years during at least eight warm interglacials.[10] While they failed to cross into North America, other forest- and plains-adapted Siberian species did, such as the long-horned Siberian bison, broad-fronted moose, woolly mammoth, grey wolf, lion, and even Siberian tiger.[11] Only once, it appears, did humans make it, which was 130.7 ± 9.4 thousand years ago, during the exceptionally warm Eemian interglacial, when sea levels rose five to nine meters higher than they are today.[12] However, these pre-Neanderthals did not survive and neither did the first wave of modern people to enter North America some 120,000 years later, while the second wave of modern humans struggled for over 1,500 years before they succeeded in entering the interior of the continent.[13] Evidence from skeletons and cave deposits indicate that they lived miserable lives.[14] Success came with the short-lived Clovis and Haskett luxury hunting cultures in North America and the Fishtail-Point Culture in South America. Success carried a high biodiversity cost.[15] Each of these cultures largely exterminated the region's mega-fauna in about five hundred years each, with some significant help in North America from the Younger Dryas Cold Spell and from the warm climates in the preceding Bølling-Allerød Interstadial during which the mega-fauna did not do so well.[16] Plants that depended for their seed dispersal on the digestive tract of the mega-fauna also went extinct.[17]

Thus, in the context of wildlife management and nature conservation in North America, it is important to note that the entry of modern humans at the beginning of the Bølling-Allerød Interstadial some 14,000 years ago marked the last time the continent's biota and landscapes were natural. Natural, in this instance, means to be influenced solely by biological evolution or Darwinian adaptation, rather than to also be affected by human culture.

## How Humans Reshaped the Landscapes of North America

What initially, and for so long, kept humans out of North America? Most likely a plethora of very large, even gigantic, and exceedingly assertive predators, and a menagerie of dangerous prey, quite unlike anything in Eurasia.[18] From the human perspective, North America's fauna was a nightmare. How does one deal with enormous, small-brained and presumably "not-so-bright," severely stressed, ever-hungry, and most determined predators, often operating in gangs?[19] Indeed, how does one deal with gigantic prey more likely to attack than flee? How could one possibly hunt, or track wounded prey, without being hunted? How could one retain possession of a kill without attracting a monstrous, hungry super-bear, huge lions, packs of big dire wolves, or fearsome saber-toothed tigers? In Pleistocene North America a diverse guild of predators dominated and shaped the fauna and the nascent ecology of human intruders to a remarkable extent.[20] There was nothing like it in the Old World, where food scarcity trumped predators as an adaptive challenge.[21]

To hold these predators in check required that humans burn enclaves for themselves and use fire as a deterrent, not unlike what Native cultures in more recent times did in California to deter grizzly bears[22] or early Europeans did to deter wolves.[23] This is also what the first Australians did some 30,000 years earlier, when they were confronted with the huge (some seven meters long) cursorial land-crocodile, *Quinkana*, the equally large, armed, and armored monitor dragon, *Varanus priscus*, an ambush-and-steal predator, giant snakes, and the small-brained marsupial lion *Thylacoleo*.[24] Seeking safety from mega-predators, people on both continents set fire liberally, and this practice left its signature in the soil deposits. In North America, as charcoal deposits increased, there was a concomitant decline in the soil of the intestinal dung bacteria, *Sporomiella*, of the mega-fauna.[25] That is, as the continent's landscapes were torched, so the mega-fauna declined, and while some carnivore genera went extinct relatively quickly, other mega-predators persisted, at least for a while.[26]

Then, some 1,500 years after their arrival, life for humans took a positive turn in North America as a short-lived luxury culture emerged in Mexico and spread rapidly into the interior and northward across the continent. This Clovis tool culture was characterized by delicate spear-points with distal grooves, and it represented a technological breakthrough. Clovis hunters killed mammoths in excess, even killing several individuals close together, and stacked the surplus meat into caches.[27] Clearly, Clovis hunters were not only capable of unprecedented hunting prowess but also no longer afraid of scavenging by mega-carnivores, or caching surplus mammoth carcasses would have been in vain. But how could this major ecological shift for human populations have occurred?

The explanation appears to be that Clovis hunters had, over the centuries, developed a miracle weapon, a deeply penetrating spear-point on a detachable fore-shaft laden with an extremely effective poison.[28] It was hurled with power and accuracy by an atl-atl. A similar system was used in Africa by the Wata people, who were traditional elephant hunters, as investigated by Ian Parker, himself a famous elephant hunter.[29] They used a long, heavy arrow, with poison from several plant species concentrated into a dosage enough to kill 70 elephants applied to the detachable arrow head. Shot with a massive long bow, the projectile was aimed low into the gut on the left side of the elephant, where it reached the small intestines. Gallons of omental fluid dissolved the poison rapidly, which was taken up by many square feet of omental and intestinal surfaces and brought into the blood stream, causing heart-failure in the elephant in minutes.

The Clovis people of North America had probably an even more potent poison, because it allowed them to kill several mammoths before they could run away and die in different localities. In fact, the suspected poison from the roots of the plant known as wolf bane or monkshood, *Aconite*, is so powerful that Aleuts used but one detachable, poisoned spear-point to kill whales. A whale so hit drowned in about twenty-four hours and drifted ashore, or beached itself in apparent agony.[30] Clearly, mega-predators could be now killed by holding them at bay with fire brands and finishing them off in minutes with poisoned spears to the gut.[31] Even the giant short-faced bear may have been killed by Clovis hunters, as its bones marked with human cuts were found at an archaeological site in Texas.[32]

The short-lived Clovis culture, which killed almost exclusively proboscidians, came to an abrupt end with the onset of a severe cold spell, the Younger Dryas, after which most of the mega-fauna was extinct. It was followed by a similar blade culture, the Folsom culture, which lasted nearly a millennium. It focused its poison projectiles on massive long-horned bison, which apparently confronted the hunters and were killed in masses. Within a millennium of heavy bison killing, the bison began to dwarf and flee hunters.[33] This made the poisoned Folsom projectile useless, and the Folsom culture, too, came to an end.

Following these ecological adjustments, all that was left of North American native species were white-tailed and black-tailed deer, caribou, mountain sheep and mountain goats, bison, pronghorns, peccaries, black bear, and coyote. A depauperate assemblage, yes, but these are, by and large, quick-learning, adaptable species that readily take advantage of humans. They were joined by Siberian immigrants, such as brown bear, grey wolf, elk, and moose, which had been in contact with early *Homo erectus* for some 1.6 million years, as well as with pre-Neanderthals, Denisovans, Neanderthals, and two waves of "out of Africa" modern humans, before colonizing North America. However, the few Siberian immigrants failed to be an adequate ecological substitute for the many extinct American mega-fauna species. This biotic depletion would lead to massive landscape changes that would challenge the human populations now well established in North America.

The extinction of mega-herbivores and large populations of lesser herbivores led to the unhindered growth of vegetation, which became fuel for lightning-set wildfires of great severity. To escape these conflagrations, humans were forced into a regime of vegetation management via fire.[34] Accordingly, fires were set not only to prevent conflagrations, but also to favor vegetation that would provide food for humans. This practice led to the development of highly sophisticated agriculture by Native Americans, which allowed the growth of dense human populations by the time America was discovered by Europeans. A similar scenario has been described for Australia.[35]

While the evidence for agricultural innovation is well recognized, the American archaeological record also reveals that existing wildlife was so severely exploited at this time that large-bodied species became rare and mainly deer and mountain sheep remained.[36] This paucity of wildlife was verified by the eye-witness reports of some of the first Europeans to enter and explore North America. Alvar Nunez Cabeza de Vaca traveled from Florida to Mexico in the years 1527–1537 and lived with Native cultures for eight years.[37] According to his reports, he encountered no alligators, wolves, or coyotes in Florida or in what became known as Louisiana. In Florida, he did see black bears and pumas, but only puma hides being used in clothing, together with mink fur. For warmth, people huddled under deer skins; no bearskins were ever in sight. Deer were present, but not plentiful, except in distant enclaves in western Tejas (or Texas now). Missing entirely were peccaries, elk, grizzly bear, coyote, grey wolf, beaver, eagle, turkeys, and even passenger pigeons, which were purportedly blotting out the sky in enormous swarms elsewhere.[38] Bison were present deep in the interior but were also heavily hunted, and utilization of game was intense, down to grinding bones for consuming.

The extreme depletion of wildlife referenced here is a classic example of Garett Hardin's tragedy of the commons.[39] The landscape was dominated by tall, old nut trees, while widespread burning was done to drive out game, or so the Spaniards assumed. During his eight years living among Native North Americans from many different tribes, Alvar Cabeza de Vaca never encountered prairie or forest fires, though lightning-scarred trees were common in the extensive, open forests of old trees. He identified nut-bearing trees and trees with medicinal properties. He was disgusted by the many fires set by Natives but did not connect the nature of the landscape with the Native fire regimes. However, even modern botanists may fail to distinguish carefully tended Native gardens from "wilderness."[40]

With the entry of Europeans to North America, their diseases began decimating native human populations.[41] Wildlife responded by multiplying and rapidly extending species' ranges.[42] Elk, peccaries, grizzly bears, and grey wolves appeared in what are today Texas, Arizona, and New Mexico.[43] Buffalo, earlier confined to the center of the continent, expanded east and west, reaching their maximum distribution by 1776 in Pennsylvania and 1812 in California.[44] Passenger pigeons, feeding on the nut and fruit trees no longer harvested by humans, exploded in numbers, their prodigious abundance an effect of a long process of habitat change by human agency.

Through this process, the abundance of wildlife that Europeans encountered in the eighteenth and nineteenth centuries led to the erroneous idea that such wildlife numbers had existed from pre-Columbian times onward, and that the North American continent had been a wilderness occupied by a small, ecologically-wise human population. Europeans did not understand that the continent had long been *civilized*, that its landscapes were extensively man-made, and that the *wilderness* they perceived was an artifact of human colonization and intrusion. The role of America's Native peoples as potent agents of landscape management has been greatly underestimated ever since.[45]

Regardless, the abundance of wildlife in nineteenth-century North America, including the re-wilding of cattle and horses as mustangs,[46] posed a problem to US and settler visions of achieving "dominion form coast to coast." That is, the abundance of buffalo, especially, supported Native peoples of exceptional health, physical development, and skill in warfare,[47] who were able to inflict painful defeats and disproportionate casualties in skirmishes with the US military. This, in turn, led to a government strategy of depriving Native tribes of sustenance to force them into surrender and submission.[48] This deliberate policy not only doomed the buffalo, but also led to a severe decline and near extinction of many other wildlife species. These effects were reinforced by provisioning large mining populations with cheap wild meat and by encouraging general uncontrolled public use of wildlife.[49] In addition, consumer markets for wildlife products grew in eastern urban centers, where luxury food and fashion items came into demand on the part of elites. The cumulative effect on wildlife was devastating and led to very serious efforts in the United States and Canada to protect the remnants and develop means and ways of recovering and spreading wildlife. The late nineteenth century was the birth hour of the North American Model of Wildlife Conservation.[50]

## Why Look at History? What Does One Learn?

As to wildlife: with the coming of humans to North America there is a great decrease in native mammals, birds, reptiles and even plants. The remaining wildlife adapts to humans. Species move from Siberia and south America into the ecological vacuum in north America. During civilizing of the continent with landscape management, horticulture and agriculture, concomitant with the growth and diversification of the human population, wildlife continues to be severely utilized and depleted. After European contact, Native American populations decline severely, allowing wildlife and "wilderness" to flourish and spread. This state of affairs is today mistaken for pre-Columbian conditions. However, wildlife abundance was again sharply curtailed in the nineteenth century in order to control Native populations, put them on reserves, and allow colonization and unhindered European settlement of the continent.[51] This was followed by a vigorous and highly successful effort in the late ninteenth and early twentieth centuries to save remaining wildlife species from extinction and multiply bird and game animal populations. This was done via a new way of conserving wildlife, known today as the North American Model of Wildlife Conservation. History, however, does not stand still: invasive mammalian species are a growing problem, while massive losses of wildlife, directly and indirectly to habitat loss and fragmentation and to diseases, remain ongoing threats.[52]

The history of North American landscapes and plants is no less dramatic. Prior to human arrival, the plant communities were closely cropped by the abundant mega-fauna, resulting in open landscapes with a paucity of trees, and a flora vigorously defended by tough and resilient wood, thorns and poisons. Fires were few. With human arrival, fires increased, mega-fauna and browsing declined, while fire hardy plant species were being selected for, and those with severe anti-herbivory defenses declined in abundance. Trees spread. This trend accelerated, until humans began steering fires so as to generate more species useful to humans. Such management also favored fire-hardy species, while overall soil fertility declined severely with the loss of the mega-fauna.[53] This civilizing of the landscape was disrupted by European colonization, followed by an increase in large, lightning caused wildfires of large size. Plant communities, freed from both severe mega-faunal browsing, as well as human burning, turned into "wilderness" perceived. There followed a large influx of Eurasian and African invasive plant species.[54] America's landscapes are not "natural," but highly artificial due to human actions, and if Sergey Zimov and his son Nikita are right, then even the tundra is an artefact of human-making, due to the extermination of mammoths, rhinos, camels, horses, etc.[55]

As to humans: the above is but a small slice of history. Contrary to a common conception, mega-faunal extinctions did not begin in North America 14,000 years ago, but were geographically widespread and began almost two million years ago with our evolution as humans.[56] There has never been an ongoing harmony between man and nature. Quite the contrary; historically, what wildlife survived, did so in spite of us, not because of us. This is the critical point. Ours was a universal, hard struggle to make the landscape safe and productive for human beings. But if so, how did it ever come to wildlife conservation? It could not be practiced by nomadic hunter gatherers who had to move when a resource ran out. It could only begin when tribes became sessile. It has something to do with ability to control wildlife for our benefit, one of the early expressions of which is domestication. Wildlife conservation, however, exercises a much less intensive control by manipulating the wild environment skillfully to favor species considered useful, interesting or entertaining. For our purposes here, it is important to ask on whose behalf were such efforts made?

In many places, such as Europe, historically, it was for the ruling elites.[57] These elites did cherish wildlife, and they ensured it survived for their benefit, enjoyment, and entertainment through a regime of extremely well-developed, professional wildlife management, based on an astonishing depth of knowledge acquired through keen observation and hands-on experience.[58] The ruling masters of these wildlife managers had visions about numbers, kind, and quality of wildlife they wanted, and they competed on that basis with their neighbors. Wildlife was a means to enhance the image of power of rulers, and conspicuous consumption was a way to demonstrate and augment prestige and position.

Regrettably, these traditions also serve as a good introduction of man's inhumanity to man, as well as the struggle of the oppressed against the arbitrariness of power. And the historic containment of power had always something to do with control over wildlife in the ownership of the elite, and between tribes in other circumstances. We must acknowledge that such realities remain relevant to conservation discussions today and, in Europe, persisted virtually in tact right up to the twentieth century.[59] Archduke Franz Ferdinand of Austria, the one whose assassination on June 28, 1914, in Sarajevo purportedly triggered World War I, amassed in his lifetime a kill of 272,511 pieces or heads of wildlife. Assuming he started killing in 1863, when ten years old, he would have killed 24 pieces or heads of wildlife per day until he was assassinated![60] But then measuring power and prowess by numbers of wildlife killed is not entirely unknown in North America and elsewhere in present times either, as revealed by wildlife law enforcement evidence.[61] In the late nineteenth and early twentieth centuries, however, the conservation movement developed not only against the backdrop of European excesses and traditions of wildlife ownership, but above all in reaction to the rapid and remorseless destruction of wildlife in North America itself. The North American movement transferred the privilege and right to hunt from the elite to all citizens, and it changed the focus from mass-killings of wildlife for the purposes of prestige and profit to selective trophy hunting. The latter was thus a means to limit the killing of wildlife.

Such was the beginning of the North American Model of Wildlife Conservation.[62] It united the brightest and the best from the United States and Canada behind a common cause, forming a Shakespearean "Band of Brothers." They included heads of states, knights (in the case of Canada), business tycoons, university presidents, senior elected politicians, artists, professors, authors, directors of national museums, and explorers. Most remarkably, Canada, then a loyal colony of Great Britain, abandoned all European policies and practices of managing wildlife and joined with the United States in a new, untested venture for conserving and managing wildlife.[63] It proved to be enormously successful, as it restored wildlife to North America and proved that the public ownership of wildlife did not lead to Garret Harding's "Tragedy of the Commons," but to a "Triumph of the Commons."[64] It fostered not only wildlife, but also new business opportunities, creating a system of generating wealth and employment by the private sector. The North American Model of Wildlife Conservation is one of the continent's great cultural achievements. However, its success depends on a functional democracy, an armed citizenry, and public acceptance of the idea of wildlife being a renewable resource for consumption.[65]

There was always a principle guiding this very American innovation; namely, that the welfare of a nation was determined not by the welfare of its elite, but the welfare of the common man. If the common man can live a life worth living, a life in which privileges are shared with the elite and rank and status are within everyone's grasp, the nation can be strong and invincible. Thus, the North American Model of Wildlife Conservation is based, in significant part, on the assumption and social acceptance of an armed citizenry. That is, all citizen have de jure and de facto access to harvesting wildlife for personal and family use, and this access can be actualized only by the ownership and habitual use of fire arms. That's one reason the North American Model, for all its successes, cannot be readily exported. Armed citizenries are not acceptable everywhere. Russians experimented with the Model after the fall of communism, but failed. Also, inherent to the Model and a nation of hunters is the ability and experience of much of its citizenry to comfortably handle firearms in an out of doors setting. And that is also a military treasure, of course, well recognized as leading to the formation in Europe of hunter regiments like the Rifle Regiments of the British tradition, the Jaegerregimente in Germany, and Chasseurs in France. These were elite units, made possible because of the hunting traditions in their countries. So, too, was the capacity of the revolutionary army formed and formulated by the hunting traditions of rural, backwoods America. Its capacities need not be debated.

One must note further in this regard, that a natural familiarity with firearms does not come from shooting small bores at stationary targets, as cadets are taught to do, but from killing a rabbit or grouse with a gun for the family table, as boys and girls did in rural America. At least one Japanese general recognized the traditional resilience of Canada and the United States derived from this cultural context when, during World War II, he opposed any invasion of North America on the grounds that there would be a gun behind every bush. There would be! And the North American Model of Wildlife Conservation is very much based on the notion that it would contribute to more than wildlife conservation. It would also contribute to the quality of life, citizenry, and nation by encouraging self-reliance and independence, as President Theodore Roosevelt many times articulated. This perspective is reflected in the founding principles and ongoing motivations of many nongovernmental organizations that are currently dedicated to preserving wildlife in abundance, such as the venerable Boone and Crockett Club, the Campfire Club of North America, Ducks Unlimited, the Wild Turkey Federation, the Rocky Mountain Elk foundation, the Wild Sheep Foundation, the Mule Deer Foundation, the Quality Deer Management Association, Whitetails Forever, the Ruffed Grouse Society, and many others.

The North American Model of Wildlife Conservation arose in the late nineteenth and early twentieth centuries via trial and error in the full brutality of the democratic system of governance in Canada and the United States. What has survived to the present is a set of policies severely tested in decades of debate in legislatures and courts and vigorous assaults in the arena of public debate. Nobody saw it as a coherent system, including wildlife historians such as Larry Jahn, which was a weakness, because wildlife managers did not see that there were principles to be vigorously defended.[66] The recognition of these came about when American agriculture turned to privatizing wildlife on its terms and the need for a clearly stated vision of what this privatization put at risk emerged.[67] Once articulated clearly, there was indeed a coherent model of policies and laws at work, protecting and fostering wildlife and one that was highly successful. It had rescued and recovered wildlife on a massive scale, was a boon to private enterprise, generating wealth and employment, and a protector of public health. And at its center was the freedom and opportunity to hunt.[68]

As in the past, the Model is continues to be under attack.[69] It is ironic that despite the great successes of the North American Model of Wildlife Conservation, and the great weaknesses of preservationist models of nature conservation, the former is now viewed by many as politically incorrect, having been formed and traditionally influenced primarily by men and by hunters. It is, therefore, relevant to ask what would happen if the Model were abandoned. Almost certainly, wildlife would become, again, the private property of the social elite, as it had been through the ages, and public land would lose much of its value and appeal to the common man. While some game species would flourish, supported by market forces, carnivores would be severely decimated if not exterminated and all migratory birds diminished. Gun ownership in North America would be severely controlled and largely eliminated. The freedom to move geographically about "natural landscapes" would be severely restricted, and concentrated, much as it is in Europe. Compared to today, the freedoms and quality of life of all North Americans would be considerably diminished. However, to understand the strengths and weaknesses of the North American Model of Wildlife Conservation, we need to place it into a broader picture of nature conservation in North America. It requires no visionary to predict severe future conflict between protectionist versus hands-on management views.

## Protectionism

The North American Model of Wildlife Conservation is one of hands-on management, striving to act in line with evidence-based planning, which applies to geographic areas where wildlife is open to the public for consumption. In the United States and Canada, these areas include all public land, except where wildlife harvest is explicitly disallowed, and some private land, at the discretion of the landowner. That is normally not the case in the very large system of national parks, nor on a significant and large segment of private land. In the late nineteenth century national parks played a crucial role protecting remnants of big game species, but did so as a matter of benign neglect, as the primary function of national parks was not necessarily conservation, but public recreation. Some parks in the United States and Canada were originally established for the benefit of large railway companies. Nature conservation was a distinctly secondary function.

With its very large size, diverse history of acquisition, and varying political directives for management, the protected areas system is so extremely complex that a highly knowledgeable insider such as Dan E. Huff can chide a group of professors critical of the system with lack of understanding.[70] Huff is quite rightly critical of policies tied to notions of "naturalness" and would like to replace "natural" with "ecological" in guidelines and management plans. He also champions a vision in which areas that contain remnants of former historical and natural biota are protected, so that the "evolution of ecosystems" may continue. However, he is apparently unaware not only that this is essentially a "back to nature" philosophy, which he wishes to dismiss, but also that ecological systems are stochastic, being shaped by positive feedback and thus subject to a continuous loss of native species and a continuous gain of invasive species, which now enjoy the protection of the park. Parks are much too small for Huff's protectionist policies. Little wonder that the national parks of the United States have been steadily loosing biodiversity, while suffering a steady increase in invasive species. There are now over 6,500 of such. And it's not a new problem.[71] Moreover, while species are indeed subject to Darwinian speciation, ecosystems are not. Ecosystems can change, but not evolve, nor can they be "healthy" or "sick," though they may be disturbed.

Regardless, and most importantly, what these totally protected areas opposed was corrective hands-on management of the biota, except in some situations where the abundance of particular species was curbed by culling; or, as in the case of Yellowstone National Park, where the local grizzly bear population that once fed on the open garbage dump and entertained visitors was eliminated after the landfill was closed. Other forms of hands-on management have included the reintroduction of wolves and the preventive burning of forests. These exceptions aside, the notion that protection of large areas from hunting preserves nature unchanged in perpetuity prevails. However, protectionism is by its nature and definition a deliberately blind, nonmanagement system, tenaciously resistant to corrective action. All biotic changes that do occur are accepted under the notion that "Nature Knows Best."

But has this been a successful approach? Perhaps this question is best answered by comparing situations where individuals on private land have attempted to reverse the impact of invasive species and enhance native species. Their experience confirms that one can only halt the steady ecological degeneration of protected areas by informed, intelligent hands-on management. Where public servants on public land hesitate, private landowners exert their will, applying evidence-based management with excellent results. And this is what makes private land so very important to North American conservation. Here we see examples of productive management unhampered by conflict between diverse interest groups, and these productive circumstances ultimately raise questions about the value and future of public lands in the grip of protectionism.

A fine example of private land conservation is Wildergarten, whose owner, Mark Vande Pol, increased the number of plant species from 60 to 360 in twenty-eight years. He succeeded in awakening to active life native and nonnative seeds dormant in the soil. Wildergarten demonstrated that nature conservation, especially the protection and support of sensitive native species cannot be achieved by blind protection, but only by knowledgeable, sophisticated application of axe, fire, and spade—that is, by targeted intervention. That's what we have to do in the future. This success came with a large gain in ecological knowledge acquired through close observation and trial and error. He defeated the persistent invasive species. However, he also showed the cost and effort required in effective rewilding.[72] Another example of hands-on management increasing biodiversity can be found in the Deseret Ranches of the Mormon Church. These achievements may be contrasted with circumstances where stunning wildlife populations on public lands placed under protectionism have withered away, turning the landscape into a wildlife desert, as exemplified by the Spatsizi Plateau Provincial Park in northern British Columbia. The cause of this is quite well known: it is that predator protection policies generate a "predator pit." In the meantime, critics grapple with environmental organizations zealously defending irrational, patently false beliefs about nature.[73] Protectionism discourages recognizing problems and acting to intercept them. After all, Nature knows best and it is not for humans to interfere! Such approaches blatantly disregard science and scholarship, and as a result, wildlife quite often suffers.

Further in this regard, national parks appear to have built-in policy approaches that inevitably lead to destruction of bears. All that is required is to have photographers and tourists following them. While they may become habituated to this, habituation is temporary, as, eventually, the animal turns about to face its followers. Bears learn well that humans are afraid of them and are emboldened. Little wonder, therefore, that in North America the number one killing ground of grizzly bears is Lake Louise in Banff National Park, and number two is in and near the Banff town site itself.[74] Railroads leading to parks leave a trail of spilled grain, attracting wildlife and leading to innumerable train collisions. Carnivores drawn to the scattered carcasses then also become victims. Habituation also leads to habituated carnivores leaving the park in hunting or trapping season and getting killed, legally, by hunters and trappers. Most significantly, the policy of not interfering can lead to wildlife extinctions as illustrated by the famous history of wolves on Isle Royal National Park, the loss of relict woodland caribou to wolves in Canada,[75] and moose in Yellowstone National Park.

## Private Lands Conservation: The Buffalo Commons Example

In contrast, private interests in North America, as noted above, can be powerful agents in the management of wildlife and in wildlife recovery. Furthermore, some of these efforts occur at very large scales. One example is the attempted rewilding of the prairies via a Buffalo Commons, conceived as "the world's largest historic preservation project and the ultimate national park," where "most of the Great Plains will become what all of the United States once was—a vast land mass, largely empty and unexploited."[76] The notion is avidly supported by the American Prairie Foundation and the Great Plains Restoration Council, as well as by the World Wildlife Fund and private entities.[77]

The Buffalo Commons' aim of returning a vast prairie region to its so-called natural state does raise many issues, though. First and foremost, the bison of today, *Bison bison*, is a dwarfed form of the large native Pleistocene bison *Bison antiquus*.[78] Dwarfing was brought about, as described earlier, by the massive kills of giant bison by Folsom hunters, which, in turn, led to selection within a millennium for flighty dwarfs. Shy, fleeing bison could not be hunted with poisoned spears, and the Folsom culture collapsed. Bison continued their shrinkage in size probably due to a replacement of nutritious C3 grasses with less nutritious C4 grasses.[79] The diminutive bison of today is thus not "natural," but a product of human making.

Further, the Bison Commons is based on the assumption that millions upon millions of bison occupied the continent from time immemorial. Archaeological data, however, as well as the observations of the first Europeans to traverse North America, show that prior to Columbus the continent was heavily peopled, and that all large mammals were relatively rare, excepting deer.[80] However, after the massive die-off of Native peoples following European entry and disease introduction, bison and other wildlife greatly increased in numbers and spread geographically. Bison increased in prodigious numbers and spread east to Pennsylvania by 1776 and California by 1812. Subsequently, bison decreased in numbers and distribution due to increased exploitation, as well as the spread of feral European horses, mustangs, till bison were virtually exterminated in the nineteenth century.[81] Millions upon millions of bison were thus an artefact of early European colonization of North America. Such abundance was not "natural."

Also to be borne in mind is the fact that creeping domestication has destroyed and continues to destroy what remains of the wild bison. It is a concern raised and articulated in exemplary fashion by Professor James Bailey in his book *American Plains Bison: Rewilding an Icon* (2013).[82] It is a concern falling on deaf ears. Very few of the surviving bison in the nineteenth century escaped being raised and held under domestic conditions. Consequently, there was from the outset of bison recovery efforts selection, deliberate and otherwise, for domestic traits. What followed added insult to injury, and has continued to the present day as most bison in public herds are treated as if they were domestic cattle on a ranch and are culled in a fashion damaging to wild bison. In his thoroughness Professor Bailey does not miss the Balayew effect, the inadvertent selection for phenodeviants by selecting for tameness.[83]

As a result of these breeding efforts, most bison today carry cattle genes, with but a very few herds excepted. Such bison appear, on principle, unacceptable for a great "naturalistic" Buffalo Commons. Fortunately, the American Prairie Foundation, which is very active in this project, as well as the Turner Foundation, appear to be aware of the hybrid problem. There is, however, a conundrum to consider. All pure-bred bison in captivity have been well subjected to domestication, whereas the bison currently in Canada's Wood Buffalo National Park, which carry cattle genes, have been so severely preyed on and reduced in numbers by wolves, that they have been "de-domesticated." They are a biological treasure! Wolves may be de-domesticating pure bison in Yellowstone National Park and in the Pink Mountain herd in northern British Columbia, as well. This is something for Buffalo Commons supporters to consider.

Worth considering also is the fact that the elk, moose, grey wolf, and grizzly bear, also proposed as part of the Buffalo Commons restoration, are not native North American Pleistocene species, but immigrants from Eastern Siberia, which spread and prospered due to the extinction of the North American mega-fauna. Human-caused extinction made the entry of Siberian species possible. Prior to human entry, none of these were able to gain a foothold in North America due to the presence of native mega fauna. They are consequently a humanly caused artefact, and not a "natural" part of North America. And, furthermore, botanically, not only is the prairie overrun by nonnative plants, but these bear no resemblance to the Pleistocene flora that originally covered North America. American grasslands are saturated with invasive exotics brought over by post-Colombian settlers to North America.[84] Fortunately, the American Prairie Foundation also appears to be well aware of this problem.

Despite these caveats, while the Buffalo Commons cannot be justified as being a recreation of pre-Columbian, let alone pre-human nature, it remains a worthwhile project if it merely aims to increase the biodiversity of native North American fauna and flora. Furthermore, its scale and vision emphatically underscore the potential for private land conservation to contribute significantly to North American wildlife efforts. As well, the intention of allowing regulated hunting on the Buffalo Commons is a positive attribute of the undertaking and in line with North American Model principles and hopefully signals a willingness to consider other warranted management interventions. A questionable practice, however, would be the removal of people from lands they have ranched for generations and making them part of the urban masses. This is in principle, of course, a criticism of all national parks, globally, if human inhabitants have been removed to make way for "Nature."[85]

In North America there are indeed huge holdings of public lands in deserts, mountains, dry prairies, taiga, tundra, and polar deserts, but, the richest and most productive lands are virtually all in private hands, mostly in the service of agriculture and forestry. These privately owned lands are crucial to North America's conservation future. There are, regrettably, huge areas under intensive use from which all wildlife is excluded, except in small parcels of wetland, homesteads, ditches, and the occasional copses and hedgerows. Thus, much progress needs to made for conservation efforts on private lands in North America. However, there are private holdings such as lands managed by the Welder Wildlife Foundation in Texas, which have been dedicated to both livestock production and wildlife conservation. Furthermore, some private holdings are variously dedicated to nature conservation, based on the personal philosophy, goals, insights, and ingenuity of their private owners. This includes giving a home to foreign endangered species. Some land holdings are supported as hunting reserves by and for the wealthy, but some are, at the same time, excellent examples of generating biodiversity. The major concern here is that a switch in land ownership can mean a turn away from applied management and sustainable use conservation practices, so that ranches and other private lands that were filled with a vibrant biodiversity become wildlife deserts, manifesting many of the problems found in protected public areas such as national parks.

Private lands thus represent both new opportunities as well as potential challenges for the North American Wildlife Conservation Model. Whether conservation efforts on such holdings will move farther toward or away from utilitarian or protectionist views will significantly decide the future of wildlife conservation in Canada and, most especially, in the United States.

## Principles of the North American Model of Wildlife Conservation

The wildlife and human history of North America and the social evolution responsible for the modern Canada and United States we perceive today are all important to the very structure of the North American Model of Wildlife Conservation and to debates concerning its future. The Model, as described, is based on seven basic policies or principles, which have been articulated in various publications and will be discussed in various chapters to follow.[86] These seven principles, which have collectively guided the North American Model's success are as follows: (1) maintain wildlife as a public trust resource; (2) prohibit deleterious commerce in dead wildlife products; (3) allocate wildlife democratically and by law; (4) ensure that all wildlife is used for legitimate purpose; (5) preserve hunting's opportunity for all; (6) recognize and manage wildlife as an international resource; and (7) ensure that science is the basis for conservation policy.

Properly applied, these principles will not only conserve wildlife and biodiversity, but also have rich economic and public health benefits.[87] Their reasonableness will also convince reasonable people who still believe in evidence-based policies, of the worth of this uniquely North American approach. Violating the Model's tenets can have serious consequences. Privatization of wildlife is but one example. If wildlife is transferred from the public trust into private ownership, the door is opened to the spread of diseases with costly consequences to society. It is, of course, important to understand that private ownership of land does not equate to private owernship of wildlife. Where wildlife exists on private land, it does not belong to the landowner. He or she may sell access rights, but not the animals themselves. Putting deer into private ownership, as in the case of game ranches, has lead to the importation and spread of bovine tuberculosis and aided the spread of chronic wasting disease, which threaten much more than the existence of wild deer populations.[88] Privatizing wildlife requires a market in wildlife product, which historically has led to massive, highly damaging, illegal killing of wildlife.

Further, when the allocation for consumption of wildlife by law is violated, such as allowing unlimited wildlife killing to certain groups in society, it can lead swiftly to extinction of wildlife through overkilling. And when the killing of wildlife becomes or appears frivolous or extends to killing beyond need or necessity, when massive wildlife is killed for thrill and bragged about on social media as with massive dove shoots in Argentina, the hunter's position as responsible steward is weakened across the board. Thrill-killing elk at a mile distance falls into the same category. The matter of how to behave as a hunter and how to treat and conserve wildlife continues to be a matter of popular and philosophical discussions, with a great many contributors, although the best known may be Theodore Roosevelt and Aldo Leopold.[89]

Fortunately, international wildlife treaties in North America have retained an even-handed allocation and effective conservation of resources such as migratory waterfowl, although there is agitation against this allocation for clearly selfish motives. The tenet that science and scholarship be the guiding light for all wildlife management is, unfortunately, honored too often in the breach rather than in the observance. Also, governments have moved away from evidence-based to opinion-based decision-making, resulting in a severe denigration of science and scholarship. One must distinguish between evidence-based scholarship and opinion-based advocacy.[90]

Finally, while there may be a nominal democracy of hunting in that all citizens in good standing may participate and the activity is governed by democratic consent, this is being undermined by legal, privileged hunting and the illegal abuse of such privilege on private lands by the elite. For these, and many other reasons, it is not only prudent to listen carefully to the critics of the North American Model of Wildlife Conservation, but to thoroughly investigate the sources of their claims, so as to distinguish between evidence-based scholarship and opinion-based advocacy, or arguments based on less than the full truth. The results can be surprising.

*NOTES*

1. Larry Littlebird, *Hunting Sacred, Everything Listens: A Pueblo Indian Man's Oral Tradition Legacy* (Santa Fe, NM: Western Edge Press, 2001).
2. A. Leopold, *Game Management* (Madison: University of Wisconsin Press, 1933).
3. D. E. Huff, "Wildlife Management in America's National Parks: Preparing for the Next Century," *George Wright Forum* 14, no. 3 (1997): 25–33.
4. T. L. Fleischer, "Ecological Costs of Livestock Grazing in Western North America," *Conservation Biology* 8, no. 3 (1994): 629–644.
5. V. Geist, S. P. Mahoney, and J. F. Organ, "Why Hunting Has Defined the North American Model of Wildlife Conservation," *Transactions of the North American Wildlife and Natural Resources Conference* 66 (2001): 175–185; S. P. Mahoney, "The North American Wildlife Conservation Model," *Bugle*, May/June 2004, 87–91; J. F. Organ, V. Geist, S. P. Mahoney, S. Williams, P. R. Krausman, G. R. Batcheller, T. A. Decker, R. Carmichael, P. Nanjappa, R. Regan, R. A. Medellin, R. Cantu, R. E. McCabe, S. Craven, G. M. Vecellio, and D. J. Decker, "The North American Model of Wildlife Conservation," *Wildlife Society Technical Review* 12, no. 4 (2012): 25.
6. I. McTaggert-Cowan, "Man, Wildlife and Conservation in North America," in *Wildlife Conservation Policy*, ed. V. Geist and I. McTaggert-Cowan (Calgary: Detselig Enterprises, 1995), 277–308.
7. J. C. Svenning, P. B. M. Peterson, C. J. Donlan, R. Ejrnaes, S. Faurby, M. Galetti, D. M. Hansen, B. Sandel, C. J. Sandom, J. W. Teborgh, and F. W. M. Vera, "Science for a Wilder Anthropocene: Synthesis and Future Directions for Trophic Rewilding Research," *Proceedings of the National Academy of Sciences* 113, no. 4 (2016): 898–906.
8. D. E. Jones, *Poison Arrows: North American Indian Hunting and Warfare* (Austin: University of Texas Press, 2009).
9. R. Barnett, B. Shapiro, I. Barnes, S. Y. Ho, J. Burger, N. Yamaguchi, T. F. Higham, H. T. Wheeler, W. Rosendahl, A.V. Sher, M. Sotnikova, T. Kuznetsova, G. F. Baryshnikov, L. D. Martin, C. R. Harington, J. A. Burns, and A. Cooper, "Phylogeography of Lions (*Pathera leo* spp.) Reveals Three Distinct Taxa and a Late Pleistocene Reduction in Genetic Diversity," *Molecular Ecology* 18, no. 8 (2009): 1668–1677.

10. A. P. Derevianko, A. V. Postnov, E. P. Rybin, V. K. Yaroslav, and S. G. Keates, "The Pleistocene Peopling of Siberia: A Review of Environmental and Behavioral Aspects," *Indo-Pacific Prehistory Association Bulletin* 25, no. 3 (2005): 57–68.
11. D. Froese, M. Stiller, P. D. Heintzman, A. V. Reyes, G. D. Zazula, A. E. R. Soares, M. Meyer, E. Hall, B. J. L. Jensen, L. J. Arnold, R. D. E. MacPhee, and B. Shapiro, "Fossil and Genomic Evidence Constrains the Timing of Bison Arrival in North America," *Proceedings of the National Academy of Sciences* 114, no. 13 (2017): 3457–3462; C. R. Harrington, "Berigian Research Notes, Giant Moose," Yukon Government Department of Tourism and Culture, June 6, 2013, http://www.tc.gov.yk.ca/publications/Moose_2007.pdf; J. Enk, A. Devault, R. Debruyne, C. E. King, T. Treangen, D. O'Roarke, S. L. Salzberg, D. Fisher, R. MacPhee, and H. Poinar, "Complete Columbian Mammoth Mitogenome Suggests Interbreeding with Wooly Mammoths," *Genome Biology* 12, no. 5 (2011), https://doi.org/10.1186/gb-2011-12-5-r51; C. R. Harrington, "Berigian Research Notes, Ancient Northern Wolves—Origins, Extinction and Replacement," Yukon Government Department of Tourism and Culture, June 6, 2013, http://www.tc.gov.yk.ca/publications/Wolf_2011.pdf; Barnett et al., "Phylogeography of Lions"; S. J. Herrington, "Subspecies and the Conservation of *Panthera tigris*: Preserving Genetic Heterogeneity," in *Tigers of the World*, ed. Ronald L. Tilson and Ulysses S. Seal (Park Ridge, NJ: William Andrew, 1988), 53–56.
12. S. R. Holen, T. A. Deméré, D. C. Fisher, R. Fullagar, J. B. Paces, G. T. Jefferson, J. M. Beeton, R. A. Cerutti, A. N. Rountrey, L. Vescera, and K. A. Holen, "A 130,000-Year-Old Archaeological Site in Southern California, USA," *Nature* 544 (2017): 479–483.
13. P. Skogland, S. Mallick, M. Cátira Bortolini, N. Chennagiri, T. Hünemeier, M. L. Petzl-Erler, F. M. Salzano, N. Patterson, and D. Reich, "Genetic Evidence for Two Founding Populations of the Americas," *Nature* 525 (2015): 104–108.
14. Glenn Hodges, "Tracking the First Americans," *National Geographic*, January 15, 2015, https://www.nationalgeographic.com/magazine/2015/01/first-americans/; D. L. Jenkins, L. G. Davis, T. W. Stafford Jr., P. F. Campos, B. Hockett, G. T. Jones, L. Scott Cummings, C. Yost, T. J. Connolly, R. M. Yoehe II, S. C. Gibbons, M. Raghavan, M. Rasmussen, J. L. A. Paijmans, M. Hofreiter, B. M. Kemp, J. L. Barta, C. Monroe, M. T. P. Gilbert, and E. Willerslev, "Clovis Age Western Stemmed Projectile Points and Human Coprolites at the Paisley Caves," *Science* 337, no. 6091 (2012): 223–228; P. Goldberg, F. Berna, and R. I. Macphaill, "Comment on DNA from Pre-Clovis Human Coprolites in Oregon, North America," *Science* 325, no. 5937 (2009): 148; A. Sistiaga, F. Berna, R. Laursen, and P. Goldberg, "Steroidal Biomarker Analysis of a 14,000 Years Old Putative Human Coprolite from Paisely Cave, Oregon," *Journal of Archaeological Science* 41 (2014): 813–817.
15. D. Duke, "Haskett Spear Weaponry and Protein-Residue Evidence of Proboscidean Hunting in the Great Salt Lake Desert, Utah," *PaleoAmerica* 1, no. 1 (2015): 109–112; Blake de Pastino, "Over 1,000 Ancient Stone Tools, Left by Great Basin Hunters, Found in Utah Desert," *Western Digs: Dispatches from the Ancient American West*, April 2, 2015, http://westerndigs.org/over-1000-ancient-stone-tools-left-by-great-basin-hunters-found-in-utah-desert/; S. Fidel, "Sudden Death: Chronology of Terminal Pleistocene Megafaunal Extinction," in *American Megafaunal Extinctions at the End of the Pleistocene*, ed. G. Hayes (New York: Springer Science & Business Media, 2008), 21–38.
16. W. S. Broecker, G. H. Denton, R. L. Edwards, H. Cheng, R. B. Alley, and A. E. Putnam, "Putting the Younger Dryas Cold Event into Context," *Quaternary Science Reviews* 29, nos. 9–10 (2010): 1078–1081; R. D. Guthrie, "Rapid Body Size Decline in Alaskan Pleistocene Horses before Extinction," *Nature* 426 (2003): 169–171; D. H. Mann, P. Groves, R. E. Reanier, B. V. Gaglioti, M. L. Kunz, and B. Shapiro, "Life and Extinction of Megafauna in the Ice-Age Arctic," *Proceedings of the National Academy of Sciences* 112, no. 46 (2015): 14301–14306; E. Pennisi, "Hot Spells Doomed the Mammoths," *Science*, July 23, 2015, http://www.sciencemag.org/news/2015/07/hot-spells-doomed-mammoths.
17. L. Kistler, L. A. Newsom, T. M. Ryan, A. C. Clarke, B. D. Smith, and G. H. Perry, "Gourds and Squashes (*Cucurbita* spp.) Adapted to Megafaunal Extinction and Ecological Anachronism through Domestication," *Proceedings of the National Academy of Sciences* 112, no. 49 (2015): 15107–15112.
18. V. Geist, "Did Predators Keep Humans Out of North America?" in *The Walking Larder: Patterns of Domestication, Pastoralism, and Predation*, ed. J. Clutton-Brock (London: Unwin Hyman, 1989), 282–294; G. C. Frison, *Survival by Hunting* (Berkeley: University of California Press, 2004); C. G. Turner, N. D. Ovadov, and O. V. Pavlova, ed., *Animal Teeth and Human Tools: A Taphonomic Odyssey in Ice Age Siberia* (Cambridge, UK: Cambridge University Press, 2013); D. Peacock, *In the Shadow of the Sabertooth* (Oakland, CA: AK Press, 2013); E. J. Neiburger, "Giant Bears Terrorize Ancient

Americans," in *Legends and Lore of Ancient America*, ed. F. Joseph (New York: The Rosen Publishing Group, 2014), 108–110.
19. B. Van Valkenburgh, M. W. Hayward, W. J. Ripple, C. Meloro, and V. L. Roth, "The Impact of Large Terrestrial Carnivores on Pleistocene Ecosystems," *Proceedings of the National Academy of Sciences* 113, no. 4 (2015): 862–867.
20. W. J. Ripple and B. Van Valkenburgh, "Linking Top-Down Forces to the Pleistocene Megafaunal Extinctions," *BioScience* 60, no. 7 (2010): 516–526.
21. V. Geist, *Deer of the World* (Mechanicsburg, PA: Stackpole Books, 1998).
22. T. Kroeber, *Ishi in Two Worlds: A Biography of the Last Wild Indian in North America* (Berkeley, CA: Berkeley Books, 1961).
23. Friedrich von Fleming, "Der Vollkommene Teutsche Jäger," *Leipzig* 1 (1719): 88, http://www.deutschestext archiv.de/fleming_jaeger01_1719/178.
24. G. H. Miller, J. W. Magee, B. J. Johnson, M. L. Fogel, N. A. Spooner, M. T. McCulloch, and L. K. Ayliffe, "Pleistocene Extinction of *Genyornis newtoni*: Human Impact on Australian Meagafauna," *Science* 283, no. 5399 (1999): 205–208; G. H. Miller, M. L. Fogel, J. W. Magee, M. K. Gagan, S. J. Clarke, and B. J. Johnson, "Ecosystem Collapse in Pleistocene Australia and a Human Role in Megafaunal Extinction," *Science* 309, no. 5732 (2005): 287–290.
25. G. S. Robinson, L. Pigott Burney, and D. A. Burney, "Landscape Paleoecology and Megafaunal Extinction in Southeastern New York State," *Ecological Monographs* 75, no. 3 (2005): 295–315; A. D. Baranosky, E. L. Lindsey, N. A. Villavicencio, E. Bostelmann, E. A. Hadley, J. Wanket, and C. R. Marshall, "Variable Impact of the Late-Quaternary Megafaunal Extinction in Causing Ecological State Shifts in North and South America," *Proceedings of the National Academy of Sciences* 113, no. 4 (2016): 856–861.
26. E. Johnson, ed., *Lubbock Lake: Late Quaternary Studies on the Southern High Plains* (College Station: Texas A&M University Press, 1987).
27. Frison, *Survival by Hunting*.
28. Jones, *Poison Arrows*; A. J. Osborn, "Paleoindians, Proboscideans, and Phytoxins: Exploring the Feasibility of Poison Hunting during the Last Glacial-Interglacial Transition," *Journal of Ethnobiology* 36, no. 4 (2016): 908–929.
29. I. Parker and M. Amin, *Ivory Crisis* (London: Chatto and Windus, 1983).
30. R. F. Heizer, "Aconite Arrow Poison in the Old and New World," *Journal of the Washington Academy of Sciences* 28, no. 8 (1938): 358–364; R. F. Heizer, "A Pacific Eskimo Invention in Whale Hunting in Historic Times," *American Anthropologist* 45, no. 1 (1943): 120–121.
31. Kroeber, *Ishi in Two Worlds*.
32. Johnson, *Lubbock Lake*.
33. M. C. Wilson, "Morphological Dating of the Late Quaternary Bison on the Great Plains," *Canadian Journal of Anthropology* 1 (1980): 81–85.
34. C. E. Kay, "Aboriginal Overkill and Burning: Implications for Modern Ecosystem Management," *Western Journal of Applied Forestry* 10, no. 4 (1995): 121–126; D. A. Burney and T. F. Flannery, "Fifty Millennia of Catastrophic Extinctions after Human Contact," *Trends in Ecology and Evolution* 20, no. 7 (2005): 395–401.
35. B. Gammage, *The Biggest Estate on Earth: How Aborigines Made Australia* (Sydney, Australia: Allen & Unwin, 2011).
36. C. E. Kay, "Aboriginal Overkill: The Role of Native Americans in Structuring Western Ecosystems," *Human Nature* 5, no. 4 (1994): 359–398; Kay, "Aboriginal Overkill and Burning"; C. E. Kay, "Aboriginal Overkill and the Biogeography of Moose in Western North America," *Alces* 33 (1997): 141–164.
37. A. N. Cabeza de Vaca, *Adventures in the Unknown Interior of America*, trans. Cyclone Covey (Lexington, KY: Cromwell-Collier Publishing, 1961); V. Geist, "Human Use of Wildlife and Landscapes in Pre-Contact Southern North America, as Recorded by Alvar Nunez Cabeza de Vaca, 1527–1536," *Beiträge zur Jagd-und Wildforschung* 43 (2018): 397–406.
38. G. G. R. Murray, A. E. R. Soares, B. J. Novak, N. K. Schaefer, J. A. Cahill, A. J. Barker, J. R. Demboski, A. Doll, R. R. Da Fonseca, T. L. Fulton, M. T. P. Gilbert, P. D. Heintzman, B. Letts, G. McIntosh, B. L. O'Connell, M. Peck, M. Pipes, E. S. Rice, K. M. Santos, A. G. Sohrweide, S. H. Vohr, R. B. Corbett-Detig, R. E. Green, and B. Shapiro, "Natural Selection Shaped the Rise and Fall of the Passenger Pigeon Genomic Diversity," *Science* 358, no. 6365 (2017): 951–964.
39. G. Hardin, "The Tragedy of the Commons," *Science* 162, no. 3859 (1968): 1243–1248.
40. E. Anderson, *Plants, Man and Life* (Boston: Little, Brown & Company, 1952).
41. H. F. Dobyns, *Their Number Become Thinned: Native American Population Dynamics in Eastern North America* (Knoxville: University of Tennessee Press, 1983); D. E. Stannard, *American Holocaust: The Conquest of a New World* (Oxford, UK: Oxford University Press, 1993).
42. W. M. Denevan, "Nature Rebounds," *Geographical Review* 106, no. 3 (2016): 1–18.
43. F. Dobie, *The Longhorns* (Boston: Little, Brown & Company, 1941); F. Dobie, *The Mustangs* (Boston: Little, Brown & Company, 1952); R. B. Gill, C. Gill,

R. Peel, and J. Vasquez, "Are Elk Native to Texas? Historical and Archaeological Evidence for the Natural Occurrence of Elk in Texas," *Journal of Big Bend Studies* 28 (2016): 1–66.

44. F. G. Roe, *The North American Buffalo: A Critical Study of the Species in Its Wild State* (Toronto: University of Toronto Press, 1951).
45. Mann et al., "Life and Extinction"; C. R. Clement, W. M. Denevan, M. J. Heckenberger, A. Braga Junqueira, E. G. Neves, W. G. Teixeira, and W. I. Woods, "The Domestication of Amazonia before the European Conquest," *Proceedings of the Royal Society B* 282, no. 1812 (2015), doi: 10.1098/rspb.2015.0813.
46. Dobie, *The Longhorns*; Dobie, *The Mustangs*.
47. J. Weston Phippen, "Kill Every Buffalo You Can! Every Buffalo Dead Is an Indian Gone," *Atlantic*, May 13, 2016, https://www.theatlantic.com/national/archive/2016/05/the-buffalo-killers/482349/; L. Cordain, "North Americans Plains Indians: Tall and Robust Meat Eaters, but Not a Milk Drinker among Them," *The Paleo Diet®*, April 22, 2016, http://thepaleodiet.com/north-american-plains-indians-tall-and-robust-meat-eaters-but-not-a-milk-drinker-among-them/.
48. S. E. Ambrose, *Crazy Horse and Custer: The Parallel Lives of Two American Warriors* (New York: Anchor, 1996).
49. P. Matthiessen, *Wildlife in America* (New York: Viking Press, 1959).
50. V. Geist, "North American Policies of Wildlife Conservation," in *Wildlife Conservation Policy*, ed. V. Geist and I. McTaggart-Cowan (Calgary: Detselig Enterprises, 1995), 75–129.
51. J. Mcdonald, "A Critique of National Parks as 'America's Best Idea,'" *New West Network Topics, Politics*, September 30, 2009, https://newwest.net/main/article/a_critique_of_national_parks_as_americas_best_idea/.
52. N. P. Snow, M. A. Jarzyna, and K. C. VerCauteren, "Interpreting and Predicting the Spread of Invasive Wild Pigs," *Journal of Applied Ecology* 54, no. 6 (2017): 2022–2032; V. Geist, D. Clausen, V. Crichton, and D. Rowledge, *The Challenge of CWD: Insidious and Dire*, Living Legacy White Paper (Calgary: Alliance for Public Wildlife, 2017), http://www.apwildlife.org/publications/.
53. V. Geist, "A Brief History of Human-Predator Conflicts and Potent Lessons," *Proceedings of the 27th Vertebrate Pest Conference*, ed. R. M. Timm and R. A. Baldwin (Davis: University of California, Davis, 2016), 3–12.
54. Anderson, *Plants, Man and Life*.
55. S. A. Zimov, "Pleistocene Park: Return of the Mammoth's Ecosystem," *Science* 308, no. 5723 (2005): 796–798.
56. V. Geist, "Disinterested Science: The Basis of Our Roosevelt Doctrine," *Fair Chase* 28, no. 4 (2013): 28–31; V. Geist, "Eine kurze Geschichte der Konflikte zwischen Menschen und Raubtieren," *Beiträge zur Jagd-und Wildforschung* 42 (2017): 63–80.
57. Fleming, "Der Vollkommene"; W. Threlfall, "Conservation and Wildlife Management in Britain," in *Wildlife Conservation Policy*, ed. V. Geist and I. McTaggart-Cowan (Calgary: Detselig Enterprises, 1995), 27–76; R. Valdez, "Exploring Our Ancient Roots. Genghis Khan to Aldo Leopold, the Origins of Wildlife Management," *Wildlife Professional* Summer (2013): 50–53.
58. Fredric II of Hohenstaufen, *The Art of Falconry*, trans. Casey A. Wood and F. Marjorie Fyfe (Stanford, CA: Stanford University Press, 1943); Fleming, "Der Vollkommene," 88; Ferdinand Freiherr von Raesfeld, *Das Deutsche Waidwerk*, 14th ed., ed. Rüdiger Schwarz (Stuttgart, Germany: Kosmos Publishers, 1980).
59. Threlfall, "Conservation," 27–74.
60. C. Oswald, *Das Rotwild der Erde: Geschichte—Verbreitung—Formenvielfalt* (Vehlefanz, Germany: Eigenverl, 2010).
61. T. Grosz, *The Thin Green Line: Outwitting Poachers, Smugglers, and Market Hunters* (Boulder, CO: Big Earth Publishing, 2004); T. Grosz, *Wildlife Wars: The Life and Times of a Fish and Game Warden* (Las Vegas, NV: Wolfpack Publishing, 2015).
62. Geist, Mahoney, and Organ, "Why Hunting Has Defined the North American Model"; D. Williams and A. Alexander, "The North American Model of Wildlife Conservation in Wyoming: Understanding It, Preserving It, and Funding Its Future," *Wyoming Law Review* 14, no. 2 (2014): 659–702.
63. V. Geist, "A Continental System of Wildlife Conservation," *Fair Chase* 15, no. 1 (2000): 15–17.
64. J. Posewitz, *Beyond Fair Chase: The Ethic and Tradition of Hunting* (Helena, MT: Falcon Guides, 1994); J. Posewitz, *Rifle in Hand: How Wild America Was Saved* (Helena, MT: Riverbend Publishing, 2004).
65. Geist, "A Brief History of Human-Predator Conflicts"; J. R. Heffelfinger, V. Geist, and W. Wishart, "The Role of Hunting in North American Wildlife Conservation," *International Journal of Environmental* Studies 70, no. 3 (2013): 399–413.
66. L. R. Jahn, Foreword to A. Leopold, *Game Management* (Madison: University of Wisconsin Press, 1986), xvii–xxx.
67. Geist, "North American Policies."
68. S. P. Mahoney and J. J. Jackson, "Enshrining Hunting as a Foundation for Conservation—The North American Model," *International Journal of Environmental Studies* 70, no. 3 (2013): 448–459.
69. M. N. Peterson and M. P. Nelson, "Why the North American Model of Wildlife Conservation Is Problem-

atic for Modern Wildlife Management," *Human Dimensions of Wildlife* 22, no. 1 (2016): 43–54.

70. F. H. Wagner, R. Foresta, R. B. Gill, D. R. McCullough, M. R. Pelton, W. F. Porter, and H. Salwasser, *Wildlife Policies in the US National Parks* (Washington, DC: Island Press, 1995).
71. Wagner et al., *Wildlife Policies.*
72. M. E. Van de Pol, "Wildergarten: A Case Study in Plant Restoration with Mark Van de Pol," lecture, Pacific Grove Museum, Pacific Grove, California, April 17, 2015.
73. A. Bergurud, "The Caribou Conservation Conundrum," in *The Real Wolf: The Science, Politics, and Economics of Coexisting with Wolves in Modern Times*, ed. T. B. Lyon and W. N. Graves (Helena, MT: Farcountry Press, 2014), chap. 7.
74. S. E. Nelson, S. Herrero, M. S. Boyce, R. D. Mace, B. Benn, M. L. Gibeau, and S. Jevons, "Modelling the Spatial Distribution of Human-Caused Grizzly Bear Mortalities in the Central Rockies Ecosystem of Canada," *Biological Conservation* 120, no. 1 (2004) 101–113.
75. Bergurud, "The Caribou Conservation Conundrum."
76. D. E. Popper and F. J. Popper, "The Great Plains: From Dust to Dust," *Planning* 53, no. 12 (1987): 12–18; D. E. Popper and F. J. Popper, "The Buffalo Commons: Its Antecedents and Their Implications," *Online Journal of Rural Research and Policy* 1, no. 6 (2006), https://doi.org/10.4148/ojrrp.v1i6.34.
77. D. Samuels, "Where the Buffalo Roam," *Mother Jones*, April 5, 2011, https://www.motherjones.com/environment/2011/02/buffalo-commons-american-prairie-foundation/.
78. Wilson, "Morphological Dating."
79. P. J. Lewis, E. Johnson, B. Buchanan, and S. E. Churchill, "The Impact of Changing Grasslands on Late Quaternary Bison of the Southern Plains," *Quaternary International* 217, nos. 1–2 (2010): 117–130.
80. Kay, "Aboriginal Overkill Biogeography of Moose"; Cabeza de Vaca, *Adventures in the Unknown Interior of America.*
81. Roe, *The North American Buffalo.*
82. J. A. Bailey, *American Plains Bison: Rewilding an Icon* (Helena, MT: Farcountry Press, 2013); J. Dupuis, "The Canadian War on Science: A long, Unexaggerated, Devastating Chronological Indictment," *Confessions of a Science Librarian*, May 20, 2013, http://confessions.scientopia.org/2013/05/20/the-canadian-war-on-science-a-long-unexaggerated-devastating-chronological-indictment/.
83. L. Trut, "Early Canid Domestication: The Farm-Fox Experiment," *American Scientist* 87, no. 2 (1999): 160–169.
84. Anderson, *Plants, Man and Life.*
85. M. Rangarajan, "The Politics of Ecology: The Debate on Wildlife and People in India, 1970–95," *Economic and Political Weekly* 31, no. 35/37 (1996): 2391–2409; M. Muhumuza and K. Balkwill, "Factors Affecting the Success of Conserving Biodiversity in National Parks: A Review of Case Studies from Africa," *International Journal of Biodiversity* 2013, http://dx.doi.org/10.1155/2013/798101.
86. Geist, "North American Policies"; Mahoney, "The North American Model"; Mahoney and Jackson, "Enshrining Hunting"; Organ et al., "The North American Model."
87. Geist, "North American Policies."
88. Geist, "North American Policies"; Geist et al., *The Challenge of CWD.*
89. Theodore Roosevelt, *The Deer Family*, ed. Theodore S. Van Dyke, Daniel Giraud Elliot, and A. J. Stone (New York: Macmillan, 1902), 1–27; C. Gordon Hewitt, *The Conservation of the Wildlife of Canada* (New York: Charles Scribner's Sons, 1921); Aldo Leopold, *A Sand County Almanac* (Oxford, UK: Oxford University Press, 1949); Jack O'Connor, *Sheep and Hunting* (New York: Winchester Press, 1974); Ted Kerasote, *Bloodties: Nature, Culture, and the Hunt* (New York: Kodansha International, 1993); Posewitz, *Beyond Fair Chase*; David Peterson, *A Hunter's Heart: Honest Essays on Blood Sport* (New York: Henry Holt, 1996); Patrick Durkin, *Deer Hunters: The Tactics, Lore, Legacy and Allure of American Deer Hunting* (Iola, WI: Krause Publications, 1997); Rob Wegner, *Legendary Deer Camps* (Iola, WI: Krause Publications, 2001); Nathan Kowalsky, ed., *Hunting—Philosophy for Everyone: In Search of the Wild Life* (Chichester, UK: Wiley-Blackwell, 2010).
90. Geist, "Disinterested Science."

John Sandlos

# The Social Context for the Emergence of the North American Model

*The earliest days of European colonization led to severe depletion of many wildlife resources (e.g., bison). In the end, it was these very excesses of slaughter that stirred public conscience, changed public attitudes, and led to the rise of conservation. Social circumstances that influenced this awakening include the disappearance of the American frontier, loss of Native American culture, and rise of European elites who began to intellectualize the fate of wildlife and other resources as a signature of national progress. This chapter will explore the various social and economic forces that enabled the rise of a conservation ethic in the wake of widespread natural resource depletion.*

## Introduction

In 1918 C. Gordon Hewitt, dominion entomologist and consulting zoologist with the Canadian Department of Agriculture, proclaimed that the previous year had represented a high water mark for the progress of wildlife conservation in Canada. In a review for the Canadian Commission of Conservation, Hewitt lauded the federal government's decisive actions to protect wildlife, first, in the creation of an interdepartmental Advisory Board on Wildlife Protection just three days before the beginning of 1917, second, in revisions to the Northwest Game Act establishing closed seasons and licensing provisions for hunting in the territorial North, and, third, in the passage the Migratory Birds Convention Act (enabling legislation for the 1916 Migratory Birds Treaty with the United States). Even though Canada had been embroiled for three years in the horrors of World War I, the federal government had retained the enthusiasm for wildlife conservation that it had displayed throughout the first two decades of the twentieth century. Among many initiatives, the Canadian federal government had infamously outflanked its US counterpart with the purchase of a privately owned bison herd from a Montana rancher in 1907; this was to be a seed herd for restocking the extirpated plains bison of the southern Canadian prairies, albeit in the fenced enclosures of Buffalo and Elk Island national parks in Alberta.[1] For good measure, between 1900 and 1920, the government added seven new national parks to the four created in the nineteenth century, many of them scenic mountain playgrounds. However, some—the two bison parks, the antelope range that became Nemiskam National Park (since deleted from the park system), and the internationally significant bird sanctuary at Point Pelee—were set aside specifically for the purposes of wildlife conservation.[2] During the same period, most provincial governments mirrored the actions of their federal counterparts, replacing the ad hoc local systems of wildlife administration with systemic and legally enforceable hunting and fishing regulations, licensing systems, and, in some cases, provincial parks or game reserves to protect vulnerable species.[3]

The United States pursued much the same expansion and consolidation of its wildlife regulatory regime. With avid sport hunter and conservationist Teddy Roosevelt sitting in the president's office from 1901 to 1909, the creation of new administrative and legal tools for wildlife protection soon rose to the top of the political agenda. Roosevelt used his powers of executive order, for instance, to establish a large system of bird refuges as a protective measure against the market hunt for plume feathers, as well as a corresponding refuge system for terrestrial wildlife. In 1909 he organized a North American Conservation Congress at the White House, which was attended by Canadian and American delegates. The US Congress also acted to protect wildlife, passing the Lacey Act in 1900, prohibiting the shipment of illegally killed wildlife across state boundaries, and the Weeks-McLean Act in 1913, granting the federal government authority to legislate waterfowl hunting seasons based on the fact that their migratory nature rendered them a form of interstate commerce. To avoid constitutional challenges to the new legislation, the United States initiated the Migratory Birds Treaty with Canada as a means to supersede state authority over wildlife regulations. As with the Canadian provinces, US state governments began to implement comprehensive game laws and licensing systems to replace the variety of local management systems that had previously dominated rural and hinterland areas.[4]

As governments in Canada and the United States established regulatory regimes to govern the hunting of wildlife, their approach mirrored the key principles of the sport hunting lobby. These included state management of wildlife as a public resource, allocation of access to wildlife for hunting through licenses and democratic regulation, severe restrictions on hunting for the market, and science as the primary knowledge base for managing wildlife. The development of this particularly North American Model of Wildlife Conservation, based as it was on seasonal limitations, bag limits, and other restrictions, in essence designated wildlife as a recreational resource, rather than a source of subsistence or income.[5] Such restrictions were crucial to the preservation, conservation, and, in some cases, the eventual restoration of key game species. At the same time, these limitations preserved some measure of democratic access to wildlife as a public resource. As many contributors to this volume argue, the contribution of sport hunters to North American wildlife conservation has been integral. In practical terms, sport hunters not only developed a philosophy and practical approach to conservation that remains in use today, but also contributed significantly to local wildlife habitat protection and umbrella initiatives such as Ducks Unlimited's relentless efforts to preserve critical wetland habitat for wildlife.[6]

On the other hand, a close examination of the social and cultural roots of the wildlife conservation movement reveals that inconsistency and outright failure sat alongside its more visionary aspirations. For instance, while North American wildlife managers hoped to increase public access to wildlife by ensuring a steady "crop" of game, the advent of game regulations throughout North America made it difficult, if not impossible, to rely on subsistence hunting, which had at least partially sustained many rural folk for generations, and indigenous people since time immemorial.[7] From the very beginning, a movement that purported to conserve democratic access to wildlife adopted the language of race and class prejudice, excoriating indigenous people, African Americans, Italians, and lower-class rural whites for their supposedly wanton hunting for the pot and the market.[8] For a movement that valued the conservation of "wild lives," the imposition of scientific wildlife management often resulted in interventions that were barely distinguishable from agricultural husbandry.[9] Perhaps most glaringly, a movement devoted to conserving wild creatures ended up killing (and in some case extirpating) thousands upon thousands of animals, as state agencies and individual sport hunters attempted to rid the countryside of the "scourge" of predators.[10]

The conservation movement invoked romantic and antimodern ideas exalting the therapeutic value of immersing oneself in the natural world (as predator in the case of the sport hunter). However, it also adopted thoroughly modern ideas of efficiency, scientific management, and state control associated with the Progressive movement of the late nineteenth and early twentieth centuries. Paradoxically, the new emphasis on scientific management meant that some wildlife herds were subject to interventions barely distinguishable from those used in agriculture. As geographer Peter Usher has argued, "Many wildlife professionals see themselves as custodians of a conservationist ethic that is above politics. This perspective sets them apart from those who manage resources such as timber, oil and gas, or minerals, although the promoter's view is not unknown in wildlife agencies when a particular resource appears to have great commercial possibilities."[11] Conservationists also contributed to new forms of wildlife commodification, as part of the tourism industry, as recreational hunting expeditions and wildlife viewing became the dominant way that North Americans interacted with wild creatures.[12] Acknowledging the complex roots and contradictory impulses of the conservation movement does not mean that we should cynically dismiss the movement's very real achievements. It may be going too far, for example, to adopt naturalist John Livingston's view that wildlife conservation is a fallacy because it reinforces human dominance of nature through its emphasis on the use-value of animals.[13] And yet, any reasonably balanced assessment of conservation history forces us recognize the movement less as a visionary impulse and more as the imperfect product of a time period when North Americans attempted to reimagine human relationships to nature in a rapidly modernizing world.

## The Origins of the Conservation Idea

The typical historical account of the wildlife conservation movement asserts that the movement was, at root, an immediate response to a profound crisis. In the late nineteenth century, a critical mass of concerned citizens, many of them sport hunters and naturalists, awoke to the plight of North American fauna in the wake of high-profile species population collapses or extinctions, most notably the severe decline of the plains bison and the complete destruction of the once numberless passenger pigeon. With the settlement of the western frontier and rapid urbanization and industrialization throughout the late nineteenth century, North Americans feared the implications of losing contact with nature (not least because of the implications this might have for the maintenance of "manliness" among North American males). Early conservation organizations such as the national-scale Boone and Crockett Club, Sierra Club, and American Bison Society, or more local groups such as Canada's Federation of Ontario Naturalists and Essex County Wild Life Association (the latter played a decisive role in the creation of Point Pelee National Park as a haven for migratory birds), worked with a growing conservation bureaucracy in government to save vital species and spaces threatened by market hunting and rapacious resource exploitation. Conservationist thought and action came too late to save some doomed species, but enough people were shocked by the scale of wildlife carnage in the latter half of the nineteenth century to conserve what was left, and this set the stage for the recovery of some species in subsequent years.[14]

There is a great deal of truth to the narrative of sudden crisis and enlightened response that is often attached to our historical understanding of wildlife conservation. But such a story also obscures much about the historical depth and complexity of the movement. Indeed, concerns about wildlife conservation and broader human relationships to the natural world were not new to the late nineteenth century. Environmental historian Richard Judd has suggested, for instance, that conservationist ideas have deep roots in the United States, particularly among eighteenth- and nineteenth-century naturalists such as John James Audubon, William and John Bartram, and Alexander Wilson. Not only did these early pioneers document some of the earliest species declines in the United States, but their writings began to develop a mixture of new (and sometimes contradictory) ideas about the natural world. These included theological and scientific ideas about natural balance and order, romantic ideas about beauty and the ontological value of primitive nature, and the notion that animals were not simply machines, but might possess complicated emotional, intellectual, and social lives similar to that of humans. They also emphasized the utility of wild animals, with ornithologists such as Wilson and Audubon noting, for example, the value of birds as destroyers of insects (an early statement of the more contemporary idea that nature should be protected because of the "ecosystem services" that it provides to humans).[15]

In Canada, natural historians such as Catherine Parr Traill, William Dawson, and Abbé Léon Provancher grappled with many of the same questions as their American counterparts, writing through the early to mid-nineteenth century of the sublime in nature and articulating a philosophy of natural theology that attempted to reconcile Darwinian evolution with the presence of divine order in the natural world. Canada's Victorian naturalists also celebrated the use-value of nature, delineating the economic value of mineral, botanical, and wildlife resources in their popular and academic writings, producing a scientific discourse that was "born of wonder and nurtured by greed," according to historian Carl Berger.[16] On a more practical level, in 1821 Hudson's Bay Company Governor George Simpson developed a beaver conservation program that included such measures as quotas, seasonal trapping restrictions, limitations on the use of steel traps, removal of trading posts, and financial incentives to trap other species. Historian Arthur Ray has speculated that the HBC program "must have been one of the earliest attempts in North America to put a primary resource industry on a sustained-yield basis."[17] Clearly the conservation movement of the late nineteenth century was not entirely spontaneous or new, but drew upon ideas and models that had been circulating for well over a century in North America.

## Conservation Goes Viral

One often overlooked reason why conservation became a popular movement by the late nineteenth century is the shifting media landscape, which offered a much wider platform for the circulation of conservation ideas. The advent of steam-powered printing presses and the subsequent explosion of newspaper and magazine publishing in the latter half of the century, along with the advent of radio in the 1920s, created unparalleled opportunities for conservationists to spread their ideas. Hunting and conservation magazines such as *Rod and Gun in Canada* (1899), *Audubon Magazine* (1887), *Forest and Stream* (1873), *Bird-Lore* (1889), and *The Ottawa Naturalist* (1887) carried conservation ideas (not to mention dire stories about the crisis of declining wildlife numbers mostly at the hands of the market hunters) into the mainstream public consciousness. The new conservation departments within government fed the growing media machine with endless press releases about parks and wildlife, or even published their own magazines and broadsheets such as the Canadian government's aptly titled *Conservation*, a monthly bulletin published by the Commission of Conservation. The proliferation of media produced an almost inevitable culture of celebrity as prolific writers on conservation issues such as John Muir, George Bird Grinnell, and William Hornaday became public figures.[18]

In Canada, the conservation celebrities were slower to emerge, but when they did, they were able to exploit rapidly expanding new media technologies. Jack Miner, a pioneer of bird banding who was dubbed the "father of conservation" by the *Minneapolis Journal*, set up a migratory bird sanctuary on his

Kingsville, Ontario, farm in 1904, attracting geese and ducks to the area by feeding them. In subsequent decades, his radio shows, lectures, books, and articles reached a large international audience, attracted by his folksy wisdom and passion for birds. Similarly, in the 1930s, the self-styled conservation prophet, Grey Owl, who railed against the depredations of the fur trade on the Canadian beaver, gained a huge international public following through articles and books, as well as on film as the star attraction of Riding Mountain National Park (and later Prince Albert National Park, along with his pet beavers, Jellyroll and Rawhide). The case of Grey Owl also illustrates some of the perils of celebrity culture: a major media scandal erupted after his death in 1938, when the press revealed he was actually an Englishman named Archie Belaney, rather than the "Indian" persona he had created to lend weight to his conservation message.[19] Though Grey Owl's credibility lay in tatters, his good friend H. U. Green, a Royal Northwest Mounted Police officer based near Riding Mountain National Park, developed a parallel career as a minor conservation celebrity under the pen name Tony Lascelles. Green's writing worked the same folksy vein as Miner: newspaper and magazine articles advocating more protections for beaver and elk, admiration for First Nations as proto-conservationists, and advocacy for the creation of a national park in the Riding Mountain area.[20] In subsequent decades, the wildlife conservation movement continued to exploit the new media landscape; the wildlife documentary, for example, became a major advocacy tool after World War II, and today the use of mobile apps, YouTube videos, and social media has become commonplace. Too often, we forget that the first flush of widespread public support for conservationism also relied on expansion of the mass media industry in the latter part of the nineteenth century.[21]

As important as mass media was in shaping and promoting the conservation agenda, there had to be an audience ready and willing to consume the movement's key messages. Popular support for wildlife conservation developed within a broader "back to nature" movement that achieved its cultural zeitgeist in North America from the 1890s to the start of World War I, though it had deeper roots and survived long past this narrow period. The back to nature movement was a direct reaction to the increasing destructiveness of industrial development. Many historians who view the movement as a cultural phenomenon have described it as a product of widespread anxiety about the destruction of nature associated with industrial development and the increasingly artificial environments that humans inhabited due to rapid urbanization. Back to nature adherents drew upon romantic ideas about the purity of the natural world and its power as a wellspring of spiritual enlightenment. However, they also drew upon what historian T. Jackson Lears has termed "antimodernism," a broad cultural movement embodied by a quest for therapeutic healing and authenticity in the face of modernity's artifice and emptiness. Antimodernism was manifest in several disaggregated forms of cultural expression, including the revival of folk crafts, the veneration of innocent childhood, and an appeal to medieval cultural values.[22]

The natural world offered a powerful symbolic and physical landscape for North Americans who sought authenticity and therapeutic healing as they engaged in primitive sport hunting expeditions, camped, climbed mountains, enrolled their kids in the Scouts program, visited national parks, or joined a local naturalist club. Some venerated the indigenous people of North America for their supposed primitiveness and presumed closeness to nature (a cultural sentiment that accounts for Grey Owl's popularity and the reason for his ruse in the first place). In 1902, the naturalist Ernest Thompson Seton founded the Woodcraft Movement, a youth organization much like the Scouts, but in which participants dressed up as "Indians" and engaged in activities supposedly representative of North American indigenous cultures. Seton's book-length treatise, *The Gospel of the Redman*, described his indigenous archetype as the "apostle of outdoor life," adding that "his example and precept are what the world needs to-day above any other ethical teaching of which I have knowledge."[23] Animals also were venerated, not only in natural history and sporting publications, but in the wildly popular "realistic" animal stories writers such as Seton and Charles G. D. Roberts began to churn out at the turn of the century. While maintaining a certain biological verisimilitude in their work (no talking pigs dressed up in jackets and hiding in brick houses are to be found in these stories), serious animal storytellers such as Seton and Roberts explored the inner motivations and desires of their animal characters, subtly suggesting how (in a post-Darwinian world) they are much like us and we much like them. For Roberts, the animal stories were a conduit along which the reader could move closer to the natural world:

> The animal story, as we now have it, is a potent emancipator. It frees us for a little while from the world of shop-worn utilities, and from the mean tenement of self of which we do well to grow weary. It helps us to return to nature, without requiring that we at the same time return to barbarism. It leads us back to the old kinship of the earth, without asking us to relinquish by way of toll any part of the wisdom of the ages, any fine essential of the "large result of time."[24]

There was considerable discord among conservationists over whether the animal stories could be considered science, with prominent naturalists John Burroughs calling them "sham" natural history and President Roosevelt decrying their authors as "nature fakers." But the simple fact that the debate reached a national stage involving the president, articles in *Atlantic Monthly* and *Everybody's Magazine*, and a widely reported tête-à-tête between Seton and Burroughs at a party thrown by Andrew Carnegie, suggests the degree to which broad issues surrounding wildlife, the human place in nature, and modernity had become a public preoccupation.[25]

For all the criticism the back to nature acolytes leveled at the modern world, they rarely questioned its fundamental precepts. While Lears argues convincingly that antimodernism was more than mere escapism, neither was it a prophetic or revolutionary movement that sought to shake the foundations of modernity in North America. In pursuing authentic and therapeutic experiences, the antimodernists sought a tonic for modernity rather than an antidote. Many of them were consummate insiders—business people, mainstream intellectuals, artists, bureaucrats, and prominent politicians such as Roosevelt—who hoped, to varying degrees, that within a rapidly changing modern world, we might maintain an integral connection to its more primitive past. Paradoxically, though, because of the mainstream nature of the movement, many expressions of antimodernism were readily co-opted and subsumed by the very modern culture it apparently abhorred. Folk arts such as music and handicrafts, for example, became marketable commodities and curiosities meant primarily for tourists, rather than representative examples of an authentic culture of the folk or alternative means of economic production.[26]

For the back to nature adherents, pursuit of authenticity in the form of primitive landscapes and wild animals became a product of the very commercial culture it sought to escape. Representations of the natural world in paintings, photographs, and the aforementioned proliferation of print all became significant means of marketing the natural world.[27] Moreover, as North Americans began to conceive of the natural world as a playground for recreation (rather than a source of subsistence), getting back to nature became a major part of a significant new industry in the late nineteenth century: tourism. Whether people traveled to exotic North American locales to take in the scenic splendor and wildlife of the national parks (including the many parks that placed wild animals in pens or fed animals as part of a nightly wildlife show), to hunt exotic animals, or to fish the best trout and salmon streams, business interests and boosters hoped to get in on the action.

From railroad companies that extended their lines to national parks in Canada and the United States to the chambers of commerce that lobbied government for road extensions and the development of national park playgrounds to tap the new market for automobile tourism, many hoped to profit. The palpable irony is that by getting "back to nature," hunters, campers, wilderness trippers, and naturalists facilitated the extension and expansion of the modern economy they purported to abhor: machine transportation, commercial development (hotels, shops, and so on), large-scale tour operations, and recreational developments such as ski resorts and golf courses.[28] Bringing people back to the wilderness meant paradoxically shaping nature to meet their expectations for comfort, scenic delight, and contact with wildlife, all of which entailed the taming (wildlife in pens, mountains with handrails) of the very quality of wildness many conservationists wanted to protect.

## Conservation and State Power

The back to nature movement embraced another deep irony: defining wild animals and landscapes as a recreational resource for public consumption required the application of modern management techniques such as scientific expertise and bureaucratic administration, as well as the imposition of state power over people and nature. If North American nature had become a public playground, then some measure of state control would have to be imposed on the mass of people who would want to go there. Over time, this meant that permit systems and visitor management plans governed access to national, provincial, and state parks; for hunters, it meant developing the previously mentioned system of licensing, seasonal restrictions, and bag limits. In some ways, this was an improvement over older patrician systems of conservation that relied on private organizations for enforcement and private hunting reserves as the cornerstone of wildlife conservation.[29] Despite the democratic appeal of the new system—in theory the public trust approach to managing parks and wildlife meant that everyone had equal access to wildlife and natural spaces—overt class and race bias featured strongly in the philosophy and practice of conservation during this period. In his influential book, *Our Vanishing Wild Life* (1913), William Hornaday referred to an "army of destruction," composed mainly of immigrants, African Americans, and "white trash," who threatened wildlife with their tendency to hunt for the pot or the market. He quoted approvingly from an article by Charles Askins asserting that "the pot-hunting negro has all the skill of the Indian, has more industry in his loafing, and kills without pity and without restraint."[30] He wrote of Italian Americans as a scourge of migratory birds:

> Let every state and province in America look out sharply for the bird-killing foreigner; for sooner or later, he will surely attack your wild life. The Italians are spreading, spreading, spreading. If you are without them to-day, to-morrow they will be around you. Meet them at the threshold with drastic laws, thoroughly enforced; for no half way measures will answer.[31]

Such sentiments reflected the broader culture of early twentieth-century nativism directed in particular at southern European and Slavic immigrants during this period. In these immigrants, conservationists found identifiable groups at which they could direct their criticism, even though people of all racial and ethnic backgrounds hunted birds during this period. However much of the concern over songbirds represented a legitimate conservation priority, public officials pursued the issue with overtly discriminatory rhetoric and policies, frequently dismissing the practice of killing songbirds as barbaric (civilized people hunted ducks, pheasants, grouse, and the like), adopting in some states discriminatory fee structures for immigrants, and in Pennsylvania creating draconian legislation in 1909 that forbade immigrants to hunt and own guns.[32]

At the same time, conservationists often harbored specific anxieties about the fact that many groups on the bottom of the racial and class hierarchy relied on hunting for the pot for supplemental income in the marketplace, activities that threatened wildlife because they were motivated by (sometimes desperate) need, and so they were difficult to control. Hornaday's book surveys many hinterland regions where wildlife could be protected, if only the demands of subsistence hunting could be eliminated in favor of the more meager desires of the sport hunter. Hornaday argued, for example, that the "real sportsmen of the world never will make the slightest perceptible impression on the caribou of Newfoundland," but "if the caribou of that Island ever are exterminated, it will be strictly by the people of Newfoundland, themselves."[33] Thus the sport hunter was free to limit the take to a small number of animals precisely because he (or, less commonly, she) was a recreationist spared the burden of necessity.

Arguably, conservationists reserved their most acerbic criticism for indigenous hunters, repeatedly calling those still largely dependent on hunting and trapping for their subsistence "improvident," "wanton," and "wasteful" hunters. In 1912, Maxwell Graham, chief of the Canadian Parks Branch's Animal Division, wrote, "They [the Indians] seem to be entirely devoid of any idea of economy in slaughtering, even though they must know they are certain to suffer from starvation, as a result of their indiscriminate waste of game."[34] At a national conference on wildlife conservation held in Ottawa in 1919, many of the assembled politicians, wildlife bureaucrats, and game officers (not to mention high-profile Americans such as Hornaday and John Burnham, president of the American Game Protective Association) agreed that indigenous hunters represented the greatest threat to Canadian wildlife. A Saskatchewan game officer, Fred Bradshaw, argued that "each year our department receives an increased number of complaints of wanton slaughter of big game by Indians."[35] Jack Miner described the aftermath of a beaver hunt in an indigenous community west of Sudbury: "The squaws were skinning the beaver. Some of them were thrown out; you never saw such a slaughter of game."[36]

From a contemporary vantage point, the accuracy of reports highlighting excessive indigenous hunting practices must be treated with caution. In northern Canada during that late nineteenth and early twentieth centuries (one of the few areas in North America where indigenous people still relied principally on hunting and gathering for their livelihood) many stories of excess slaughter relied on the testimony of non-Native hunters and trappers (especially those who flooded into northern Canada when fur prices rose after World War I), who were only too happy to draw the attention of the authorities to their indigenous counterparts. In other cases, sport hunters and northern explorers clearly denigrated the hunting practices of their indigenous guides as a way to exalt and validate the nobility of their own destruction of wildlife (a common trope in sport hunting narratives from around the globe).[37] Ironically, scattered reports from police and missionaries suggest that northern indigenous hunters were equally contemptuous of their non-indigenous counterparts, critical of their overly long trap-lines, contemptuous of their use of poison and aircraft, and hostile to the sport hunter's penchant for procuring trophies rather than food.[38] Certainly there are well documented cases of overkill and wasteful hunting (sometimes for the market) by indigenous hunters in North America; to ignore them risks engaging in the same romantic stereotypes of the ecological Indian that Seton and other promoted.[39] And yet, conservationists often adopted the more extreme view that all indigenous hunting people and communities were incapable of self-regulation, and that wasteful hunting was part of their racial character. Although this imagery stands in stark contrast to the primitive authenticity the back to nature advocates ascribed to indigenous people, Seton and like-minded authors were quick to point out that the contemporary native was a fallen version of the precontact archetype, tainted by alcohol, poverty, disease, and the adoption of modern hunting weapons. They argued that, when it came to hunting, savagery lurked alongside an essence of ecological nobility. In an account of his journey to the remote wood bison range of northern Alberta, Seton wrote that "the mania for killing that is seen in many white men is evidently a relic of savagery, for all of these Indians and half-breeds are full of it." He went on to suggest that, for the Cree, Chipewyan, and Métis hunters on the northern bison range, "it is nothing but kill, kill, kill every living thing they meet."[40]

The conservation measures taken to combat such apparent improvidence often carried severe consequences for local hunters. Indigenous people and other "squatters" who lived on lands designated as parks or sanctuaries in Canada and the United States were routinely expelled, more often than not with no compensation in the form of cash or new land.[41] In one egregious case, park wardens at Riding Mountain National Park aggressively removed a small community of the Keeseekoowenin Ojibway First Nation from the park in 1936, burning the houses and barns to the ground.[42] Park authorities excluded indigenous and nonindigenous communities from many other iconic mountain spaces and smaller, more obscure national, provincial, and state parks in eastern North America, a practice that remained routine until a more consultative approach was adopted with indigenous people in Alaska and northern Canada in the 1970s.[43] The reasons for expelling people varied from place to place, but authorities cited wildlife protection as much as they did maintaining the "wilderness" and "playground" character of park landscapes.[44]

Game regulations also dramatically impacted subsistence-oriented communities in rural and hinterland areas, curtailing access to wild meat through bag limits and seasonal restrictions. In Canada's Northwest Territories, for example, the Northwest Game Act of 1917 caused great anxiety among indigenous hunters because it imposed unheard-of seasonal restrictions on species that had always been freely hunted.[45] That same year, the Migratory Birds Convention Act imposed continent-wide seasonal restrictions that allowed hunting of waterfowl to begin only on September 1, well after many birds had left the high northern latitudes.[46] The consequences for breaking the game regulations were hardly trivial: between 1936 and 1944 the Canadian government sent thirteen indigenous hunters to jail for terms ranging from thirty days to six months, often with hard labor included in the sentence, for killing bison in Wood Buffalo National Park.[47] Undoubtedly, many of the species that were the target of wildlife legislation required protection, but from the perspective of rural folk and indigenous hunters, the introduction of game regulations and the creation of protected areas sometimes seemed harsh and arbitrary. Moreover, the lack of consultation with those affected very likely increased incidences of poaching in remote areas that were difficult to police, and where local people killed animals, in part, as a form of resistance to state conservation initiatives.[48]

In addition to regulating humans, the early conservation movement applied a great deal of its energies to the intensive management of wild animals. Like proponents in other renewable resource sectors, wildlife conservationists were steeped in the Progressive-Era idea that science and bureaucratic expertise could control nature in such a way as to maximize ecological productivity for human uses. In forestry, this meant fire suppression and the creation of even-aged tree plantations; for the fisheries stocking programs, it meant management based on maximum sustainable yields.[49] Although wildlife was not generally thought of as a primarily commercial resource, in 1933 Aldo Leopold famously summed up the basic philosophy of wildlife management as "the art of making land produce sustained crops of wild game."[50] If it seemed paradoxical to harvest animals that were at essence wild and subject to a range of ecological influences, early conservationists zealously tried to enhance the population of desirable species—mostly ungulates—for the use of recreational hunters.

The primary means of enhancing wild game herds was an extensive and widespread program of predator control. The job of killing wolves, coyotes, mountain lions, and, sometimes, bears had been carried out informally for decades by any hunter, trapper, or rancher with a gun on his or her shoulder, dogs to let loose, or a strychnine bait to lay out (the enticing meat ironically taken from wild ungulates). Such antipathy toward problem predators was certainly understandable among economically marginal herders or those who relied on game animals for food, but the ferocity of the attack on predatory animals veered toward irrational destruction by the late nineteenth century. Between 1850 and 1880, wolf bounties and the price of pelts resulted in the extirpation of the species from most parts of its former range. By the twentieth century, wolf control programs continued in Canada, but in the United States, the growing conservation bureaucracy applied science and technology to the anticoyote campaign. The US Bureau of Biological Survey (BBS) developed a research wing on animal control and developed partnerships with university researchers. One result was the M-44 "coyote getter," a cyanide gun that predator control agents partially buried, hoping it would explode in the animal's face when the animal disturbed a carefully laid bait. After World War I, new poisons, such as compound 1080 and thallium, provided a more deadly and selective means to attack canids. The scale of the killing efforts is staggering: while there are no reliable data for wolves, mountain lions, or grizzly bears, each of these species disappeared from much of their range in the lower forty-eight states; the BBS killed 1,884,897 coyotes between 1915 and 1947 (with little effect on the species as it engages in compensatory breeding and has adapted well to anthropogenic environments). Thousands of "nontarget" scavenger birds and small mammals likely died from picking at poisoned bait carcasses. Despite the lack of records to provide precise numbers, the available evidence suggests that that the scale of the killing too often went beyond mere control (however one may feel about the environmental ethic surrounding such a concept) and veered toward the very wanton destruction so often decried within the conservation movement.[51]

One could object that the impetus for predator control came as much from large-scale sheep and cattle interests as it did from sport hunters and other wildlife enthusiasts. Yet, prior to World War II, wildlife conservationists engaged in the war on varmints as enthusiastically as any rancher. Game wardens routinely killed predators in Canadian and US national, state, and provincial parks (where there were no cattle to protect), a practice that continued almost habitually in Canada until the 1970s, long after it had been abandoned in many other jurisdictions.[52] William Hornaday celebrated the destruction of predatory birds and mammals as an essential measure for game protection; in Canada, Jack Miner devoted large sections of his writing to excoriations of "cannibal birds" such as crows and various raptors, and he killed them in large numbers to protect the more morally upstanding geese and ducks at his Kingsville sanctuary.[53] Ernest Thompson Seton served as a predator control officer in New Mexico, where he wrote a semifictional story about his (somewhat regretful) killing of the wolf Lobo, an "outlaw" and the so-called King of the Currumpaw.[54] In his treatise on wildlife conservation, C. Gordon Hewitt argued against extermination of predators because they sometimes played a useful role controlling agricultural pests, but "a degree of control must be exercised to prevent such an increase in numbers as would affect the abundance of the non-predatory species."[55] However dedicated conservationists may have been to saving some types of North American wildlife, they also devoted a great deal of effort toward the destruction of species deemed undesirable.

Wildlife conservationists certainly were not unanimous on this issue. Major figures such as Muir and Grinnell opposed the practice; major doubts emerged in the mid-1920s, when the population of the deer herd on the Kaibab Plateau above the Grand Canyon increased dramatically in the wake of predator control programs, though reductions in grazing by domesticates and bans on human hunting in the area likely also played a role. A major debate erupted over the issue at the American Society of Mammalogists meeting in 1924, with some participants leveling sharp criticism at senior BBS zoologist Edward Goldman's strong antipredator position.[56] In the 1930s and 1940s, zoologists such as Paul Errington, Adolph Murie, and his brother Olaus began to produce landmark studies suggesting that carrying capacity, rather than predation, was the determining factor for prey species.[57] By the late 1940s Aldo Leopold had famously done an about-face on his game cropping theories, writing in his essay "Thinking like a Mountain" about how his act of killing a wolf as a young man had failed to account for the role the animal played regulating the population of grazing animals, and his own mournful regret watching the death of the "fierce green fire" in its eyes.[58] Such objections eventually curbed the worst excesses of the predator control programs, paving the way for such bold and successful programs as the Yellowstone wolf reintroduction in 1995.[59] Predator control is still used controversially in some jurisdictions when prey species are thought to be threatened, and some hunters are still prone to shoot coyotes, foxes, and other varmints on sight in jurisdictions where it is legal and/or encouraged through a bounty system.

The early wildlife conservation movement also imposed intensive managerial control over game animals. Once again following Progressive-Era doctrines about efficient production, wildlife managers at times engaged in intensive herd management schemes barely distinguishable from those used by ranchers. Conservationists enacted particularly intrusive management programs for bison (despite their symbolic representation as icons of the western wilderness), placing most of the remaining animals on fenced-in reserves or on private ranches, where they were husbanded as domestic animals. Canada was an early pioneer in this approach, positioning itself as a world leader in bison conservation when, as mentioned previously, it purchased one of the last remaining herds of plains bison from Montana ranchers Michael Pablo and Charles Allard in 1907. Between 1909 and 1914, the Canadian government funded the transfer of the 700-strong herd to the newly created Buffalo National Park in Wainwright, Alberta, and Elk Island National Park near Edmonton. These parks were the first to be fenced in Canada, in part to separate the bison from surrounding farms, but also to present an appealing and accessible wildlife menagerie for tourists (one that also included moose and elk). The Wainwright bison even starred in a Hollywood western, the 1926 version of *The Last Frontier*.[60]

By the early 1920s, however, the publicity surrounding Canada's efforts to save the last remaining bison had descended into controversy. Faced with the problem of managing an exploding bison population on a limited range, the chief of the Parks Branch's Animal Division, Maxwell Graham, promoted a scheme whereby the surplus bison population at Wainwright could be transferred to the much larger and under-stocked range in the more northerly and newly created Wood Buffalo National Park. Zoologists from around the world decried the plans due to the risks of both hybridization between plains bison and the wood bison subspecies[61] and transmission of bovine tuberculosis from the Wainwright herds to the wood bison. Graham steadfastly defended the program, however, which was likely a reflection of a broader split between "hands-on" wildlife managers (Graham was educated at the agricultural college in Guelph, Ontario) and university-based zoologists steeped in ecologist Frederic Clements's ideas about noninterference with the balance of nature. From 1925 to 1928, the Parks Branch transferred 6,673 plains bison by train and barge to a new range in northern Canada, where they spread tuberculosis and brucellosis to the bison herds. In 1925, an article by "A Canadian Zoologist" (likely William Rowan at the University of Alberta) denounced the transfer program in no uncertain terms: "Never before in the annals of conservation, as far as we are aware, have the last survivors of a unique race of animals been knowingly obliterated by a department of conservation."[62]

One of the more bizarre aspects of early conservation discourse is the way some early advocates expanded their notions of managerial production to embrace the commercialization of even the rarest wildlife species. C. Gordon Hewitt wrote in his 1921 volume on wildlife conservation that "the greatest value of the buffalo . . . lies in the possibility of its domestication." He also enthusiastically supported breeding experiments to cross cattle and bison to create an ideal commercial beef animal (to so-called beefalo or cattalo, a few of which were produced in government labs, none of which were fertile). Hewitt wrote passionately of the need to protect the declining muskox population in the Arctic, but also enthusiastically about the commercial potential associated with domesticating the species.[63] Hewitt's position as dominion entomologist within the Department of Agriculture may explain some of his seemingly conflicted positions. Within the Parks Branch, however, Maxwell Graham also maintained enthusiasm for the commercial potential of bison crossbreeds:

> Looking to the future success of the experimental cross-breeding between buffalo and domestic bovines, it is imperative that a reserve stock of pure blood bison of the highest potency should be kept in reserve, so that the ultimate fixed type of new range animal may continue to pass on to successive generations the prepotent qualities of the true bison, hardiness, thriftiness, a valuable robe, and first-class beef qualities.[64]

In 1919, Canadian Parks Commissioner James Harkin sat on a Royal Commission to investigate the potential of a domestic reindeer and muskox industry as an agricultural base for economic development in northern Canada. It is possible that the other two members of the commission, a railway commissioner and the manager of an abattoir, pushed through the commission's recommendations to introduce European reindeer to northern Canada and create pilot projects for the domestication of caribou and muskox.[65] Yet, Harkin was also among the members of the more resolutely conservationist interdepartmental Advisory Board on Wildlife Protection that advocated for the translocation of a small muskox herd from Ellesmere Island to research stations investigating the potential for commercial production in Alaska and the west coast of Hudson Bay.[66] Very few of these projects came to fruition due to the logistical challenges of operating in the Arctic, but for the contemporary observer it remains startling how much the rhetoric of wildlife conservation became conflated with the language of commercial development in the early decades of the twentieth century. Nor was such enthusiasm for wildlife domestication and commercial production unique to Canadian conservationists. Historian Andrew Isenberg has demonstrated that commercial wildlife boosterism was also common in the United States, where the vast majority of bison descended from the few that escaped depredations of market hunters ended up "conserved" on commercial ranches, the source of those high-end bison burgers that one can sometimes find in grocery store freezers. Indeed, North Americans have never mustered the political will to set aside a large portion of unfenced prairie land for an ecologically viable herd of wild bison. The Buffalo Commons project described in the previous chapter and the recent reintroduction of the animals to Banff National Park suggest some progress toward the restoration of wild bison herds in North America. The slow pace and controversy surrounding such projects is nevertheless a testament to the limitations of historic efforts to save one of North America's icons of wildness.[67]

Commercial production and predator control sit at the extreme end of a range of interventions that wildlife managers imposed on wildlife in the twentieth century. As Clementsian ideas of ecological balance began to fade in the interwar years, ecologists such as Charles Elton, Eugene Odum, and Arthur Tansley developed new models of nature as sites of dynamic energy flow with sudden fluxes or declines in populations dependent on changes in the carrying capacity of the range. Under the new paradigm, wildlife herds might be cropped to control disease and overgrazing or restocked to better utilize undergrazed rangelands.[68] In Wood Buffalo National Park, wardens began to cull the formerly endangered wood bison in a futile effort to control tuberculosis, a program that expanded into full-blown disease control and commercial production as two small abattoirs operated in the park throughout much of the 1950s and 1960s. The Parks Branch also developed a commercial culling program for bison at Elk Island National Park to keep the population in check on its fenced and predator-free range.[69] Historian Alan MacEachern has written about the Parks Branch's penchant between 1910 and the late 1960s for gifting "excess" wildlife to zoos, university scientists, natural history museums, the occasional game ranch, and the Walt Disney company (for one of its many wildlife films). The Parks Branch donated some of these animals (sometimes demanding cost recovery) or traded them for exotics to display in several of the national park zoos, but MacEachern argues that donations to foreign countries served the purpose of cultural diplomacy by ensuring typically Canadian fauna were represented on foreign soil. The practice ceased in the 1970s, due primarily to concerns over cost and wildlife mortality during shipping, but the Parks Branch's half-century of faunal donations provides perhaps the clearest evidence that wildlife conservation was as firmly wedded to human priorities as it was to saving species simply because they were irreplaceable.[70]

## Conclusion

On a spring morning during a backcountry hike in Yellowstone National Park, I crawled out of my unexpectedly snow-covered tent and took in the new light of the day. Across a meadow adjacent, a pronghorn antelope came in full gallop, close enough that I could hear the animal exhale with exertion. Speed, motion, and grace coalesced in an experience of such quality and rarity that it has been repeated in my life only a few other times during encounters with bison and caribou. No doubt I owe—all of us owe—such powerful (albeit brief) moments of contact with a wild animal to the conservationists who had the foresight to protect the pronghorn and other species threatened with overhunting and habitat loss. Environmental historians have done much over the past forty years to acknowledge this collective debt, producing a vast array of works that often celebrates the achievements of conservation pioneers in North America. Not the least of their many achievements was the development of a particularly North American Model of Wildlife Conservation based on balancing public access to game animals with sound management practices through state regulation and scientific direction.

It is equally important to acknowledge that the mainstream ideas of the wildlife conservation movement were as much products of their time as they were a visionary agenda. With roots dating back to the eighteenth century, wildlife conservation reached its apex in response to the rapid territorial expansion and industrialization that placed intense pressure on North American fauna in the latter half of the nineteenth century. In spite of the conservationists' early successes, recent histories have engaged in a more critical appraisal of the movement and have been less prone to celebrate a pantheon of conservation heroes and more likely to point to the many contradictions at the core of efforts to save dwindling animal populations. If conservationists condemned improvident and wasteful hunting, historians have indicated that they often framed their critiques in terms of race and class prejudices that likely exaggerated the depredations of Hornaday's "army of destruction" and underestimated the impacts of the sport hunting fraternity. At the same time, while conservationists may have decried market hunting and promoted the dissemination of public access to wildlife through state regulation, they were not immune (especially those in the halls of government) from the commercialism that was the norm in other resource sectors, dreaming up new ways to manage and exploit wildlife as a tourist attraction or as ranch animals. Pointing to such idiosyncrasies in the North American conservation movement does not require that we cynically dismiss its achievements. We should not, on the other hand, ignore the fact that the early wildlife conservation movement was at times selectively democratic and almost thoroughly grounded in modern notions of managerialism, production, and utility. Even if some early conservationists understood wild animals and wild nature as an almost transcendent antidote to modern life North America, for the most part, the conservation movement was an expression of exactly those modern values. Such a historical perspective reminds us that conservation was not (and is not) a politically neutral science or a value-free management practice but was deeply embedded in the complicated social and political world it simultaneously sought to shape and change.

*NOTES*

1. C. Gordon Hewitt, "Conservation of Wild Life in Canada," *Nature* 99 (1917): 246–247.
2. Janet Foster, *Working for Wildlife: The Beginnings of Preservation in Canada*, 2nd ed. (Toronto: University of Toronto Press, 1998); C. Gordon Hewitt, "The Coming Back of the Bison," *Natural History* 19, no. 6 (1919): 553–565.
3. T. Loo, *States of Nature: Conserving Canada's Wildlife in the Twentieth Century* (Vancouver: University of British Columbia Press, 2006).
4. For overviews of the US wildlife policy during this early period, see K. Dorsey, *The Dawn of Conservation Diplomacy: US-Canadian Wildlife Protection Treaties in the Progressive Era* (Seattle: University of Washington Press, 1998); T. R. Dunlap, *Saving America's Wildlife: Ecology and the American Mind, 1850–1990* (Princeton, NJ: Princeton University Press, 1988); J. Sandlos, "Nature's Nations: The Shared History of Game, Fish, and Forest Protection between Canada and the United States," *International Journal of Environmental Studies* 70, no. 3 (2013): 358–371.
5. V. Geist, "North American Policies of Wildlife Conservation," in *Wildlife Conservation Policy*, ed. V. Geist and I. McTaggart Cowan (Calgary: Detselig Enterprises, 1995), 75–129; S. P. Mahoney and J. J. Jackson, "Enshrining Hunting as a Foundation for Conservation—The North American Model," *International Journal of Environmental Studies* 70, no. 3 (2013): 448–459. For an overview of the development of wildlife law, see J. Tober, *Who Owns the Wildlife? The Political Economy of Conservation in Nineteenth-Century America* (Westport, CT: Greenwood Press, 1981).
6. For an overview of the sport hunting movement's contribution to conservation, see J. F. Reiger, *American Sportsmen and the Origins of Conservation*, 3rd ed. (Corvallis, OR: Oregon State University Press, 2001). For prairie waterfowl conservation, see S. Stunden Bower, *Wet Prairie: People, Land and Water in Agricultural Manitoba* (Vancouver: UBC Press, 2011); A. S. Hawkins, R. C. Hanson, H. K. Nelson, and H. M. Reeves, eds., *Flyways: Pioneering Waterfowl Management in North America* (Washington, DC: US Fish and Wildlife Service, 1984).
7. T. Loo, "Making a Modern Wilderness: Conserving Wildlife in Twentieth-Century Canada," *Canadian Historical Review* 82, no. 1 (2001): 92–121.
8. K. Jacoby, *Crimes Against Nature: Squatters, Poachers, Thieves and the Hidden History of American Conservation* (Berkeley: University of California Press, 2001); K. Jacoby, "Class and Environmental History: Lessons from the 'War in the Adirondacks,'" *Environmental History* 2, no. 3 (1997): 324–342; B. Parenteau, "Care, Control and Supervision: Native People in the Canadian Atlantic Salmon Fishery, 1867–1900," *Canadian Historical Review* 79, no. 1 (1998): 1–35; B. Parenteau, "A 'Very Determined Opposition to the Law': Conservation, Angling Leases, and Social Conflict in the Canadian Atlantic Salmon Fishery, 1867–1914," *Environmental History* 9, no. 3 (2004): 436–463; L. Warren, *The Hunter's Game: Poachers and Conservationists in Twentieth Century America* (New Haven, CT: Yale University Press, 1997).
9. J. Sandlos, *Hunters at the Margin: Native People and Wildlife Conservation in the Northwest Territories* (Vancouver: University of British Columbia Press, 2007).
10. D. Worster, *Nature's Economy: A History of Ecological Ideas*, 2nd ed. (Cambridge, UK: Cambridge University Press, 1994), 258–290.
11. P. J. Usher, "Property Rights: The Basis of Wildlife Management," *National and Regional Interests in the North: Third National Workshop on People, Resources and the Environment North of 60*, June 1–3, 1993 (Ottawa: Canadian Arctic Resources Committee, 1984), 389–390.
12. G. Gillespie, *Hunting for Empire: Narratives of Sport in Rupert's Land, 1840–70* (Vancouver: University of British Columbia Press, 2008); T. Loo, "Of Moose and Men: Hunting for Masculinities in British Columbia, 1880–1939," *Western Historical Quarterly* 32 (2001): 296–319; A. MacEachern, *Natural Selections: National Parks in Atlantic Canada, 1935–1970* (Montreal:

McGill–Queen's University Press, 2001); J. Sandlos, "Nature's Playgrounds: The Parks Branch and Tourism Promotion in the National Parks, 1911–1929," in *A Century of Parks Canada, 1911–2011*, ed. Claire Campbell (Calgary: University of Calgary Press, 2011), 53–78.

13. J. A. Livingston, *The Fallacy of Wildlife Conservation* (Toronto: McClelland and Stewart, 1981).
14. For works that celebrate various aspects of the early conservation movement, see J. Foster, *Working for Wildlife*; Reiger, *American Sportsmen*; J. Trefethan, *An American Crusade for Wildlife* (New York: Boone and Crockett Club, 1975); R. Nash, *Wilderness and the American Mind*, 3rd ed. (New Haven, CT: Yale University Press, 1982); M. Oelschlaeger, *The Idea of Wilderness from Prehistory to the Age of Ecology* (New Haven, CT: Yale University Press, 1991); and C. Merchant, "Women of the Progressive Conservation Movement: 1900–1916," *Environmental Review* 8, no. 1 (1984): 57–85.
15. R. Judd, "A 'Wonderfull Order and Ballance': Natural History and the Beginning of Forest Conservation in America, 1730–1830," *Environmental History* 11, no. 1 (2006), 8–36; R. Judd, *The Untilled Garden: Natural History and the Spirit of Conservation in America, 1740–1840* (Cambridge, UK: Cambridge University Press, 2009).
16. C. Berger, *Science, God, and Nature in Victorian Canada: The 1982 Joanne Goodman Lectures* (Toronto: University of Toronto Press, 1983); S. Glickman, *The Picturesque and the Sublime: A Poetics of the Canadian Landscape* (Montreal: McGill–Queen's University Press, 2000); S. Zeller, *Inventing Canada: Early Victorian Science and the Idea of a Transcontinental Nation* (Toronto: University of Toronto Press, 1987).
17. A. Ray, "Some Conservation Schemes of the Hudson's Bay Company, 1821–50: An Examination of the Problems of Resource Management in the Fur Trade," *Journal of Historical Geography* 1, no. 1 (1975): 49–68.
18. The celebrity status of Hornaday and Grinnell is obvious to anyone who spends time with conservation magazines and books from this period, as they are constantly quoted or featured as writers. There are many biographies of John Muir that deal with aspects of his celebrity; for a good overview, see D. Worster, *A Passion for Nature: The Life of John Muir* (Oxford, UK: Oxford University Press, 2008).
19. J. Miner, *Jack Miner and the Birds—Some Things I Know about Nature* (Toronto: Reilly and Lee, 1923); Grey Owl, *The Collected Works of Grey Owl* (Toronto: Prospero Books, 2004); Loo, *States of Nature*, chap. 3.
20. The author has examined a large collection of Green's writings, published and unpublished, in the following collection: Whyte Museum of the Canadian Rockies, H.U. Green Papers, M2 43.
21. For a broad overview of celebrity environmentalism, see D. Brockington, "Powerful Environmentalisms: Conservation, Celebrity and Capitalism," *Media, Culture, and Society* 30, no. 4 (2008): 551–568; S. Sullivan, "Conservation Is Sexy! What Makes This So, and What Does This Make? An Engagement with Celebrity and the Environment," *Conservation and Society* 9, no. 4 (2011): 334–345.
22. T. J. Jackson Lears, *No Place of Grace: Antimodernism and the Transformation of American Culture, 1880–1920* (New York: Pantheon, 1981). For a discussion of antimodernism in Canada, see essays by I. McKay, L. Jessup, G. Moray, and R. B. Phillips in L. L. Jessup, ed., *Antimodernism and Artistic Experience: Policing the Boundaries of Modernity* (Toronto: University of Toronto Press, 2001); I. McKay, *Quest of the Folk, CLS Edition: Antimodernism and Cultural Selection in Twentieth-Century Nova Scotia* (Montreal: McGill–Queen's University Press, 2009). For the back to nature movement, see G. Altmeyer, "Three Ideas of Nature in Canada, 1893–1914," in *Consuming Canada: Readings in Environmental History*, ed. C. Gaffield and P. Gaffield (Toronto: Copp Clark, 1995), 98–105. For the connections between American romanticism and the wilderness movement, see W. Cronon, "The Trouble with Wilderness: Or, Getting Back to the Wrong Nature," *Environmental History* 1, no. 1 (1996): 7–28.
23. Ernest Thompson Seton, *The Gospel of the Redman: An Indian Bible* (London: Psychic Press, 1970); Ernest Thompson Seton, *Two Little Savages; Being the Adventures of Two Boys Who Lived as Indians and What They Learned* (New York: Grosset & Dunlap, 1911). For more on the mainstream urge to imitate indigenous people during this period, see S. E. Bird, *Dressing in Feathers: The Construction of the Indian in American Popular Culture* (Boulder, CO: Westview, 1996); Joe Sawchuck, ed., *Images of the Indian: Portrayals of Native People*, Readings in Aboriginal Studies, vol. 4 (Brandon, Manitoba: Bearpaw Publishing, 1995); Daniel Francis, *The Imaginary Indian: The Image of the Indian in Canadian Culture* (Vancouver: Arsenal, 1993).
24. Charles G. D. Roberts, *Kindred of the Wild: A Book of Animal Life* (New York: Thomas Nelson and Sons, 1900), 19.
25. For more animal stories, see Charles G. D Roberts, *The Heart of the Ancient Wood* (Toronto: McClelland and Stewart, 1974); Ernest Thompson Seton, *Lives of the Hunted* (Toronto: George N. Morang, 1901); Ernest Thompson Seton, *Wild Animals I Have Known* (Toronto: McClelland and Stewart, 1977). For the "nature fakers"

controversy, see John Burroughs, "Real and Sham Natural History," *Atlantic Monthly* 91 (1903): 298–309; Theodore Roosevelt, "Nature Fakers," *Everybody's Magazine* (September 1907): 427–430. For discussions, see Alec Lucas, "Nature Writers and the Animal Story," in *A Literary History of Canada*, Vol. 1, ed. Carl F. Klinck (Toronto: University of Toronto Press, 1965), 380–404; R. McDonald, "The Revolt against Instinct: The Animal Stories of Seton and Roberts," *Canadian Literature* 84 (1980): 18–29; J. Sandlos, "From within Fur and Feathers: Animals in Canadian Literature," *Topia: A Canadian Journal of Cultural Studies* 4 (Fall 2000): 73–91.

26. McKay, *Quest of the Folk*.
27. J. K. Cronin, *Manufacturing National Park Nature: Photography, Ecology, and the Wilderness Industry of Jasper* (Vancouver: University of British Columbia Press, 2011); L. Jessup, "The Group of Seven and the Tourist Landscape in Western Canada, or the More Things Change," *Journal of Canadian Studies* 37, no. 1 (2002): 144–179; L. Jessup, "Landscapes and Sport, Landscapes of Exclusion: The 'Sportsman's Paradise in late 19th-Century Canadian Painting," *Journal of Canadian Studies* 40, no. 1 (200): 71–124.
28. H. K. Rothman, *Devil's Bargains: Tourism in the Twentieth-Century American West* (Lawrence: University of Kansas Press, 1998); J. Urry, *The Tourist Gaze* (London: Sage, 2002). For general studies of wilderness and tourism, see D. Louter, *Windshield Wilderness: Cars, Roads and Nature in Washington's National Parks* (Seattle: University of Washington Press, 2006); Chris J. Magoc, *Yellowstone: The Creation and Selling of an American Landscape: 1870–1903* (Albuquerque: University of New Mexico Press, 1999); Gabrielle Zezulka-Mailloux, "Laying the Tracks for Tourism," in *Culturing Wilderness in Jasper National Park: Studies in Two Centuries of Human History in the Upper Athabasca River Watershed*, ed. I. S. MacLaren (Edmonton: University of Alberta Press, 2007), 233–259; C. J. Taylor, "The Changing Habitat of Jasper Tourism," in *Culturing Wilderness in Jasper National Park: Studies in Two Centuries of Human History in the Upper Athabasca River Watershed*, ed. I. S. MacLaren (Edmonton: University of Alberta Press, 2007), 199–225.
29. Karl Jacoby, *Crimes against Nature*; Parenteau, "A 'Very Determined Opposition'"; Warren, *The Hunter's Game*. One exception to this pattern is Quebec, where an older neo-European patrician approach placed private sport hunting groups at the center of conservation practice until World War One. See D. Ingram, *Wildlife Conservation and Conflict in Quebec, 1840–1914* (Vancouver: University of British Columbia Press, 2013).
30. W. Hornaday, *Our Vanishing Wild Life: Its Extermination and Preservation* (New York: New York Zoological Society, 1913), 95. For a specific study of conservation and African Americans, see S. E. Giltner, *Hunting and Fishing in the New South: Black Labor and White Leisure after the Civil War* (Baltimore, MD: Johns Hopkins University Press, 2010).
31. Hornaday, *Our Vanishing Wild Life*, 88.
32. Warren, *The Hunter's Game*; A. Rome, "Nature Wars, Culture Wars: Immigration and Environmental Reform in the Progressive Era," *Environmental History* 13, no. 3 (2008): 432–453; G. E. Allen, "Culling the Herd: Eugenics and the Conservation Movement in the United States, 1900–1940," *Journal of the History of Biology* 46, no. 1 (2013): 31–74.
33. Hornaday, *Our Vanishing Wild Life*, 146.
34. M. Graham, Chief of the Animal Division, National Parks Branch to James Harkin, Director, National Parks Branch, December 7, 1912. RG 85, vol. 665, file 3911, pt. 1, Library and Archives Canada (hereafter LAC).
35. Commission of Conservation of Canada, *National Conference on Conservation of Game, Fur-Bearing Animals and Other Wild Life* (Ottawa: King's Printer, 1919), 22.
36. Commission of Conservation of Canada, *National Conference on Conservation*, 29.
37. J. Sandlos, *Hunters at the Margin*, 149–152; C. Campbell, "A Genealogy of the Concept of 'Wanton Slaughter' in Canadian Wildlife Biology," in *Cultivating Arctic Landscapes: Knowing and Managing Animals in the Circumpolar North*, ed. D. G. Anderson and M. Nuttall (New York: Berghan, 2004), 154–171.
38. For a discussion, see Kerry Abel, *Drum Songs: Glimpses of Dene History* (Montreal: McGill–Queen's University Press, 2005), chap. 7.
39. S. Krech, *The Ecological Indian: Myth and History* (New York: W. W. Norton, 2000); S. Krech, "Reflections on Conservation, Sustainability, and Environmentalism in Indigenous North America," *American Anthropologist* 107, no. 1 (2005): 78–86.
40. Ernest Thompson Seton, *The Arctic Prairies* (New York: Harper and Row, 1911), 20, 174.
41. P. Burnham, *Indian Country, God's Country: Native Americans and the National Parks* (Washington, DC: Island Press, 2000); Jacoby, *Crimes Against Nature*; R. H. Keller and M. F. Turek, *American Indians and National Parks* (Tucson: University of Arizona Press, 1998); Jean Manore, "Contested Terrains of Space and Place: Hunting and the Landscape Known as Algonquin Park, 1890–1950," in *The Culture of Hunting in Canada*, ed. Jean Manore and Dale Miner (Vancouver: University of British Columbia Press, 2007), 121–147; J. Sandlos, "Federal Spaces, Local Conflicts: National Parks and

the Exclusionary Politics of the Conservation Movement in Ontario, 1900–1935," *Journal of the Canadian Historical Association* 16 (2005): 293–318; M. D. Spence, *Dispossessing the Wilderness: Indian Removal and the Making of the National Parks* (Oxford, UK: Oxford University Press, 1999).

42. J. Sandlos, "Not Wanted in the Boundary: The Expulsion of the Keeseekowenin Ojibway Band from Riding Mountain National Park," *Canadian Historical Review* 89, no. 2 (2008): 189–221.
43. T. Catton, *Inhabited Wilderness: Indians, Eskimos and National Parks in Alaska* (Albequerque: University of New Mexico Press, 1997); B. Martin, "Negotiating a Partnership of Interests: Inuvialuit Land Claims and the Establishment of Northern Yukon (Ivvavik) National Park," in *A Century of Parks Canada, 1911–2011*, ed. Claire Campbell (Calgary: University of Calgary Press, 2011), 273–301; J. Sandlos, "National Parks in the Canadian North: Co-Management or Colonialism Revisited?" in *Indigenous Peoples, National Parks, and Protected Areas: A New Paradigm Linking Conservation, Culture, and Rights*, ed. Stan Stevens (Tucson: University of Arizona Press, 2014), 133–149.
44. T. Binnema and M. Niemi, "'Let the Line be Drawn Now': Wilderness, Conservation and the Exclusion of Aboriginal People from Banff National Park in Canada," *Environmental History* 11 (October 2006): 724–750.
45. Sandlos, *Hunters at the Margin.*
46. D. Gottesman, "Native Hunting and the Migratory Birds Convention Act: Historical, Political, and Ideological Perspectives," *Journal of Canadian Studies* 18, no. 3 (1983): 67–89.
47. W. H. Routledge, Patrol Report, Fort Saskatchewan to Fort Simpson, Report of the Royal Northwest Mounted Police, Sessional Papers No. 15 (1899), RG 85, vol. 769, file 5222, pt. 1-2, LAC; Maxwell Graham, *Canada's Wild Buffalo: Observation in the Wood Buffalo Park* (Ottawa: Department of the Interior, 1923), p. 11; Records of arrests and convictions for violating game regulations in areas under Canadian federal jurisdiction were found at the Library and Archives Canada, specifically Record Group 10, vol. 8409, file 191/ 20-14-1, pt. 1, LAC; RG 85, vol. 1214, file 400-2-3, pt. 3.
48. Jacoby, *Crimes against Nature*; Sandlos, *Hunters at the Margin.*
49. S. Hays, *Conservation and the Gospel of Efficiency: The Progressive Conservation Movement, 1890–1920* (New York: Atheneum, 1959).
50. A. Leopold, *Game Management* (New York: Charles Scribner's Sons, 1933), 3.
51. Worster, *Nature's Economy*, 258–290. For a broad history of changing attitudes to wolves, see K. Jones, *Wolf Mountains: A History of Wolves along the Great Divide* (Calgary: University of Calgary Press, 2002); K. Jones, "Never Cry Wolf: Science, Sentiment, and the Literary Rehabilitation of *Canis lupus*," *Canadian Historical Review* 84, no. 1 (2003): 65–94.
52. A. MacEachern, "Rationality and Rationalization in Canadian National Parks Predator Policy," in *Consuming Canada: Readings in Environmental History*, ed. Pam Gaffield and Chad Gaffield (Toronto: Copp Clark, 1995), 197–212.
53. Hornaday devotes the entirety of chapter 8 of *Our Vanishing Wild Life* to the subject of predators. Miner's work, *Jack Miner and the Birds*, is rife with commentary on the immoral nature of predatory birds and the methods the author uses to destroy them.
54. Ernest Thompson Seton, "Lobo, King of the Currumpaw," in *Wild Animals I Have Known* (Toronto: McClelland and Stewart, 2009), 3–29.
55. C. Gordon Hewitt, *The Conservation of the Wild Life of Canada* (New York: Charles Scribner's and Sons, 1921), 193.
56. T. R. Dunlap, "'The Coyote Itself—Ecologists and the Value of Predators" *Environmental Review* 7, no. 1 (1983): 54–70; T. R. Dunlap, "Values for Varmints: Predator Control and Environmental Ideas," *Pacific Historical Review* 53, no. 2 (1984): 141–161.
57. P. Errington, "Vulnerability of Bobwhite Populations to Predation," *Ecology* 15, no. 2 (1934): 110–127; A. Murie, *Ecology of the Coyote in Yellowstone*, Fauna Series No. 4 (Washington, DC: US Department of the Interior, National Parks Service 1940); O. Murie, *The Wolves of Mt. McKinley*, Fauna Series No. 5 (Washington, DC: Department of the Interior, 1944).
58. A. Leopold, *A Sand County Almanac and Sketches Here and There* (London: Oxford University Press, 1949).
59. Much of the scientific literature on the Yellowstone wolf reintroduction supports the idea of a trophic cascade, whereby the restoration of a top predator has impacts that ripple through many levels of the food chain. See W. J. Ripple and R. L. Beschta, "Trophic Cascades in Yellowstone: The First 15 Years after Wolf Reintroduction," *Biological Conservation* 145, no. 1 (2012): 205–213; J. S. Halofsky, W. J. Ripple, and R. L. Beschta, "Recoupling Fire and Aspen Recruitment after Wolf Reintroduction in Yellowstone National Park, USA," *Forest Ecology and Management* 256, no. 5 (2008): 1004–1008. For a dissenting view on the trophic cascade theory, see A. D. Middleton, Matthew J. Kauffman, Douglas E. McWhirter, Michael D. Jimenez, Rachel C. Cook, John G. Cook, Shannon E. Albeke, Hall Sawyer, and P. J. White, "Linking Anti-Predator Behaviour to Prey Demography Reveals Limited Risk

Effects of an Actively Hunting Large Carnivore," *Ecology Letters* 16, no. 8 (2013): 1023–1030.

60. J. Brower, *Lost Tracks: Buffalo National Park, 1909–1939* (Edmonton: Athabasca University Press, 2008); Foster, *Working for Wildlife*; Hewitt, "The Coming Back of the Bison."
61. Whether or not the wood bison is actually a subspecies of the plains bison has been hotly debated for decades. While the Canadian government has designated the wood bison a threatened species, and wildlife NGOs readily distinguish between plains and wood bison for the purposes of conservation, the scientific community contains dissenting views. See V. Geist, "Phantom Subspecies: The Wood Bison *Bison bison 'athabascae'* Rhoads 1897 Is Not a Valid Taxon, but an Ecotype," *Arctic* 44, no. 4 (1991): 283–300; M. A. Cronin, M. D. MacNeil, N. Vu, V. Leesburg, H. D. Blackburn, and J. N. Derr, "Genetic Variation and Differentiation of Bison (*Bison bison*) Subspecies and Cattle (*Bos taurus*) Breeds and Subspecies," *Journal of Heredity* 104, no. 4 (2013): 500–509.
62. A Canadian Zoologist, "The Passing of the Wood Bison," *Canadian Forum*, July 1925, 301–305. For overviews, see J. Brower, *Lost Tracks*; J. Sandlos, "Where the Scientists Roam: Ecology, Management and Bison in Northern Canada," *Journal of Canadian Studies* 37, no. 2 (2002): 93–129; Sandlos, *Hunters at the Margin*.
63. Hewitt, *Conservation of the Wild Life*, 136–142.
64. M. Graham, *Canada's Wild Buffalo: Observation in the Wood Buffalo Park* (Ottawa: Department of the Interior, 1923), 12.
65. J. Rutherford, J. McLean, and J. Harkin, *Report of the Royal Commission to Investigate the Possibilities of the Reindeer and Musk-Ox Industries in the Arctic and Sub-Arctic Regions of Canada* (Ottawa: The King's Printer, 1922).
66. Minutes of the Advisory Board on Wildlife Protection, February 28, 1916. RG 10, vol. 4084, file 496,658, LAC.
67. A. C. Isenberg, *The Destruction of the Bison, An Environmental History, 1750–1920* (Cambridge, UK: Cambridge University Press, 2000).
68. Worster, *Nature's Economy*, 291–314.
69. Loo, *States of Nature*; Sandlos, *Hunters at the Margin*; Sandlos, "Where the Scientists Roam."
70. A. MacEachern, "Lost in Shipping: Canadian National Parks and the International Donation of Wildlife," in *Method and Meaning in Canadian Environmental History*, ed. Alan MacEachern and William Turkel (Toronto: Nelson, 2009), 196–213.

# 4 The Great Early Champions

JAMES PEEK

*As people began to understand that North American wildlife species and other natural resources were not inexhaustible and were in peril, conservation leaders began to emerge. These individuals used their talents and leadership abilities to champion the developing North American conservation ethic, founding many of the structured organizations that would have enormous influence on North America's conservation movement and the establishment of the Model. This chapter explores the contributions of individuals such as Theodore Roosevelt, George Bird Grinnell, Gifford Pinchot, C. Gordon Hewitt, William Hornaday, Clifford Sifton, and Wilfred Laurier, whose achievements remain a benchmark for, and prime example of, conservation professionalism. It also provides some historical perspective on the formation of organized public efforts for conservation.*

## Introduction

It has been said that many ardent conservationists start out being inherently curious about nature, that interest being enhanced by handling skins and bones of wildlife specimens, often obtained by shooting. This practice develops into sport hunting as an occupation that consumes much time and thought. Then, confirmed hunters move on to appreciate fish and wildlife resources in their natural entirety. Whether this progression is sufficiently the case to be the typical example or not isn't as important as the knowledge that the work of several individuals with exceptional talent and leadership who pioneered and championed the North American conservation ethic generally followed this pattern.

North America's hunting heritage originated with the Old World traditions of preserving lands for the gentry to hunt at the exclusion of the common man. However, the needs of the common man to feed his family, and the egalitarian interests of the early organized hunters in North America, began to diverge from the European traditions in the late 1800s and early 1900s and move towards more modern approaches to hunting and management of game species. The New York Association for the Protection of Game was highly influential in these initial efforts, campaigning to eliminate the killing of game out of season.[1] That association was composed of well-to-do New York hunters, including Robert B. Roosevelt, Theodore Roosevelt's father.

## Theodore Roosevelt

As Theodore Roosevelt (1858–1919) made his "Rough Rider" presence felt on the conservation scene, the hunter and the conservationist became forever entwined on this continent. Roosevelt had asthma as a youngster, and his father encouraged outdoor experience in part as a means to address his asthmatic condition.[2] Young Roosevelt developed a strong interest in wild birds, going on to create a small museum with specimens he collected. Collections such as Roosevelt's, and the extensive one that William T. Hornaday created for the US National Museum, were a primary means of obtaining avian specimens for identification and preserving their skins. Roosevelt was a serious student of the written materials on birds and became highly respected as an authority himself on identification and taxonomy of the native species. He became a highly skilled and passionate birder, just as he would, eventually, become an ardent hunter. His philosophy thus became highly complex in regard to nature and its use by humans, and his actions as president reflect this well.

While earlier presidents had roles in designating public lands for various conservation and management purposes, Roosevelt was the primary progenitor of the modern arrangement of national forests, national parks, and wildlife refuges in the United States. He greatly expanded the national forest system to 150 million acres and established sixteen national parks and fifty-one national bird reserves that are now part of the national refuge system. His actions were often highly protectionist in nature, while at the same time he also hunted across the globe and indeed has often been remembered and described as a wilderness hunter.[3]

Roosevelt was a strong supporter of using the best science to understand and conserve the wildlife resource, due to the influence of his father. However, following John James Audubon, George Bird Grinnell, and Frank M. Chapman of the Audubon Society, he also considered the aesthetic values of wildlife to be equally important. As president, he appointed Gifford Pinchot to head the new Division of Forestry in the Department of Interior, which eventually became the US Forest Service, now in the Department of Agriculture. Pinchot was well known for his espousal of the philosophy of wise use: forests (and by implication other natural resources) were to be used "for the greatest good, for the greatest number, in the long run." This philosophy was accepted by Roosevelt as the most useful way to manage forests, in spite of opposition from friends such as John Muir and W. T. Hornaday. It was this philosophy that developed first into the sustained-yield and then the sustainable-use approaches to resource management that prevail today. The original wise-use policy emphasized that all citizens are to be considered and are important to the North American Conservation Model.

## George Bird Grinnell

Perhaps the best example of Roosevelt's combined dedication to fair-chase hunting and conservation is his co-founding of the Boone and Crockett Club, with George Bird Grinnell, in 1887. The pair enlisted other well-to-do hunters who espoused the fair-chase hunting philosophy and opposed market killing of wildlife to join them. The club was dedicated primarily to the preservation and increase of big game mammals and lobbied extensively for laws that would encourage wildlife conservation and better enforcement, generally, and in the new Yellowstone National Park and California parks, especially. This all took place at a time when excessive exploitation of wildlife was rampant, when populations of many species were at low levels, and when some even became extinct (Merriam's elk, Audubon bighorn, heath hen, passenger pigeon). While Roosevelt may be the most well-known supporter of fair-chase hunting, other individuals were also highly influential during his time, none more so than George Bird Grinnell (1848–1938). Grinnell was the founder of the first state Audubon Society in New York and was a highly important mentor for Roosevelt.[4] He co-founded the American Game Association, a forerunner of the Wildlife Management Institute. From 1876 to 1911 he edited *Forest and Stream* magazine, which gave him a forum for reporting on steadily dwindling wildlife, and most especially bison. He was also a champion of protecting Yellowstone National Park, which Roosevelt supported. His broad and varied contributions to wildlife conservation, including his concern for nonhunted species, his writing about enforcement of game laws, and his support for sustained-yield forest management and habitat conservation remain vital components of the North American Model today. In many ways, Grinnell's philosophy was the foundation for development of the Model itself.

Grinnell hunted in the vicinity of what is now Glacier National Park, beginning in 1885, and became influential in establishing the park. He also became a strong advocate for preserving the few remaining bison in Yellowstone and encouraged the US Department of Interior to find animals outside of the park for protection as well. He was an outspoken advocate for Native Americans, including the Pawnee, Gros-Ventre, and Cheyenne, and worked for reasonable treaties for the tribes. Grinnell, like Roosevelt, had a complex philosophy concerning natural resources and was a major spokesman for their proper management and preservation. He is another example of the fair-chase hunter who contributed enormously in the early years to the inception of the North American Conservation Model.

## Gifford Pinchot

Gifford Pinchot (1865–1946) is primarily recognized as an advocate for sustained-yield forestry. However, forests provide habitats for a wide array of wildlife species, and their proper management and conservation are important components of the North American Model. Pinchot was educated at Yale before there was a forestry degree and went to France to study the subject after graduating.[5] He became involved early on in forest management and conservation issues and caught Roosevelt's eye as a logical choice for managing the federal forests.[6] He became a major advisor to Roosevelt during his presidency, as well as a personal friend. He was a member of the Boone and Crockett Club at Roosevelt's urging, and the first chief of the US Forest Service after it was transferred to the US Department of Agriculture in 1905. Pinchot is credited with popularizing the concept of conservation of natural resources through multiple use and sustained yield among the general public. He was also an advocate of using science to manage forests and other resources, as were most of the early supporters of fair-chase hunting, and he was a founder of the Society of American Foresters. Pinchot was fired in 1910 by President Taft, after a controversy over coal claim issues in Alaska. He was subsequently twice elected governor of Pennsylvania (1922–1927, 1931–1935), where he worked to have rural roads built and maintained.

## William T. Hornaday

William T. Hornaday (1854–1937), founder of the US National Museum in 1882 (now part of the Smithsonian Institution), might be best known for his extensive collections of native birds and mammals for the museum and subsequently for the New York Zoological Park. Hornaday was active during the period when there were extensive efforts to identify and catalogue all wildlife species in North America. This meant that skeletons, furs, and feathers were needed, which entailed forays into the country and the collection of many specimens. These collections could be extensive and include rare species, which caused criticism from some in the conservation sector. However, after Hornaday was sent into central Montana to collect a specimen from a remnant herd of bison, he became a major advocate for preservation of the few remaining individuals and is given credit for preventing the extinction of bison through his efforts.[7] Subsequently, Hornaday wrote *Our Vanishing Wildlife* in 1913, a documentation of the rapid reduction and loss of North American species through excessive killing, which helped galvanize interest in protecting fish and wildlife resources. He also founded a permanent Wild Life Protection Fund in 1913 to help support his lobbying efforts. Hornaday was initially a sport hunter, as well as a collector for study, although he came to resist hunting after becoming involved in the bison and passenger pigeon debacles. He strongly advocated elimination of the devastating fur seal slaughters, arguing during a bitter, five-year campaign for their protection and the creation of preserves for their Pribilof Island breeding grounds. He ceased to hunt entirely after this experience. Roosevelt aligned himself with Hornaday in trying to protect the seals, with the creation of the Seal Treaty of 1911. The treaty involved the United States, Britain (on behalf of Canada), Russia, and Japan, and mandated a ban on killing fur seals and sea otters in American waters by Congress a year later. Hornaday's commitment to wildlife was profound and personal, and he could be highly obstinate and disagreeable when at odds with others.[8]

### George Shiras III

George Shiras III (1859–1942) is known for his wildlife photography and for the discovery of a subspecies of moose inhabiting the United States and adjacent Canada, which was named the Shiras moose after him. He was born in Pennsylvania, attended Cornell and then the Yale Law College, graduating in 1883. He was elected to the Pennsylvania State House of Representatives in 1889 and 1890, and then to the Fifty-Eighth Congress in 1903–1905. As a child, he developed an interest in the outdoors and nature and first became an avid hunter before switching to wildlife photography in his early teens. When *National Geographic* magazine ran seventy-four of his photographs in 1906, interest in the magazine was stirred, and this began the extensive use of photographs in articles featured in that publication today. As a congressman, he wrote the first legislation that authorized the Migratory Bird Law, which finally passed in 1913. He developed a friendship with Theodore Roosevelt and became a member of the Boone and Crockett Club in 1906.

### Charles Sheldon

Charles Sheldon (1867–1928) is considered the father of Denali National Park.[9] Raised in a family engaged in marble quarrying and manufacturing, he graduated from Yale in 1890. He then worked in the family business, for railroads in Michigan and Chihuahua, and was involved in mining in Mexico. He was able to retire from business in 1903 and began an association with the Biological Survey in 1904. His friends included Theodore Roosevelt, George Bird Grinnell, and Gifford Pinchot, and he was elected to the Boone and Crockett Club in 1905. He spent long periods of the 1903–1908 years in the upper Yukon and the Denali Park region hunting brown bears, studying the wildlife, and collecting specimens for the National Museum. He also studied desert sheep in Arizona and Sonora from 1912 to 1922. During his 1906 expedition to Denali, he decided to support an effort to establish a game refuge or park in the area. He realized that market hunting needed to be stopped and was instrumental in getting it banned. While he was an avid sheep hunter, he strenuously lobbied the US Congress to establish a national park in Denali. His connections with other conservation advocates, including Roosevelt, meant that his views would be heard. His efforts were supported by the Boone and Crockett Club, the American Game Protective Association, and the Camp Fire Club. Finally, in 1916, Judge Wickersham, Alaska's delegate to Congress, and Senator Key Pittman of Nevada introduced the legislation to designate the park, which was passed in early 1917. President Woodrow Wilson signed the act into law in February 1917. Sheldon's book, *The Wilderness of Denali* (1930) was published posthumously by his wife with the help of the Biological Survey.

### Herbert L. Stoddard

Herbert L. Stoddard (1889–1970) is recognized as one of the most important early wildlife biologists in the southern United States.[10] Stoddard was born in Illinois and educated himself as a naturalist. He worked as a taxidermist in the Milwaukee Public Museum and the Field Museum of Natural History in Chicago from 1910 to 1924 and was then hired by the Bureau of Biological Survey to study bobwhite quail in the Georgia-Florida Red Hills area. He helped to establish the profession of wildlife management with Aldo Leopold and became the foremost authority on bobwhite quail and forest management in the longleaf pine–wiregrass forests of the south. He practiced the use of fire to replenish the productivity of these forests and to maintain them as quail habitat. He established the still-existing Tall Timbers Research Station in 1958 on land donated by Henry Beadel. The Komarek brothers, Ed and Roy, were also involved and carried on Stoddard's efforts when he passed away. The Fire Ecology Conferences that Stoddard's efforts launched in 1962 have, ever since, carried information about the ecology and role of fire in many wildlife habitats across the world. His 1931 publication, *The Bobwhite Quail: Its Habits, Preservation and Increase*, was among the first comprehensive investigations in the field of wildlife management. Stoddard's attention to habitat preferences and forestry practices as a way of minimizing predation and increasing prey species remains a valuable approach in wildlife habitat management today.

## Aldo Leopold

Aldo Leopold (1887–1948) first became interested in wild birds as a youngster, when living along the Mississippi River in Burlington, Iowa.[11] Birds were abundant and diverse along the river system. With the support of his father, he started hunting at an early age, primarily for birds. Leopold was trained as a forester, one of "Pinchot's boys," forestry being a new profession that appealed to him. In 1909, he graduated from Yale, where the Pinchot family had established an endowment to train graduate foresters. He was hired by the US Forest Service and was sent to the Carson National Forest in New Mexico, where he eventually became the supervisor. It was apparent early on that his interests did not lie in the retail forestry sector but rather involved the wider dynamics of natural systems.

Aldo Leopold's definition of game management was utilitarian in nature, coinciding with the philosophy of Roosevelt and Pinchot.[12] He believed the restoration of wildlife involved the creative use of the same tools that had, up until that time in North America, helped destroy it—the axe, plow, cow, fire, and gun. Despite his sensitivities to nature, he argued that simply trying to restore wildlife by controlling only the gun had not been very successful and that a broader, hands-on habitat-based approach was needed. He pointed out that the legal status of wildlife being owned by the state (government) dated back to Roman law, which concluded that since wild animals had no individual owner, the resource should be considered property of the state until taken into personal possession. He also, therefore, frequently discussed the rights of private landowners in controlling trespass onto their property and supported their right to control wildlife existing on their lands. Leopold believed that hunting's future would be controlled by three factors: laws, incentives based on self-interest, and personal ethics. From the self-interest perspective he referred to the need to allow private landowners to be compensated for allowing hunters access to their land, but also referred to courtesies extended between hunters over where and when to hunt and making efforts to shoot and prepare game in ways that retained meat, bones, and hide in the best condition.

Leopold portrayed wildlife associated with managed forests and farmland as a crop to be managed, much as trees and corn, with the exceptions of waterfowl and wilderness game. These species he described as best managed as a by-product rather than a primary product of the land, along with timber and agricultural crops. Leopold also recognized that some species did not lend themselves to being managed as crops, while others most certainly did. He categorized grizzly bears, elk, moose, caribou, and mountain sheep as examples of wilderness game that did not thrive in managed settings. Today, this categorization may be viewed as outdated since forest management can obviously be directed at providing habitat, particularly for elk and moose, and reclaimed mine settings in Alberta have been used very successfully by mountain sheep. Regardless, those old categories of farm game, forest and range game, waterfowl, and wilderness still have utility when it comes to acquiring habitats to benefit various species. We aren't going to manage grizzly bears, elk, and moose as farm game, but white-tailed deer occupy agricultural lands, and on refuges established for waterfowl, intensive land management and manipulation of water levels are commonplace.

In his *Sand County Almanac* (1949), Leopold expanded on these views and emphasized that humans should consider themselves part of a community including all living things and that we need to develop an ecological conscience that recognizes this. This final writing, published posthumously, illustrated Leopold's development from a bird watcher to a hunter, to a wildlife biologist, and, finally, to a conservation visionary whose evolving views were eventually captured in his land ethic philosophy.

Leopold was involved throughout his career in the creation of the national wilderness system in the United States, but others made significant contributions as well. While Leopold urged the creation of a wilderness in the Gila River region in 1924, Arthur Carhart, a landscape architect employed by the US Forest Service, had recommended wilderness-style management of what is now the Flat Tops Wilderness in Colorado in 1920 and subsequently the Boundary Waters Canoe Area in Minnesota, and the two men had discussed wilderness creation before that. Bob Marshall, Howard Zahniser, and Ernie Olberholtzer were all instrumental in urging Congress to formally recognize wilderness, with Zahniser writing the first drafts of what would become the 1964 Wilderness Act.

## Arthur Carhart

Arthur Carhart (1892–1978) was highly influential, sometimes behind the scenes, and presaged the concept of the North American Wildlife Conservation Model and some of its current challenges, in his reminder to the deer hunter.

> I wanted you to see your stake in the deer herds of North America, your property rights in them and in the public hunting grounds where, if we guard our rights, you and I may continue to hunt. You, and the other stockholders, through your game and fish officials, may hope and plan for future hunting; you must guard that future so this American tradition may persist, not only for you and me, but for others to come.[13]

While Carhart started his career in the US Forest Service, he became a well-known advocate for wilderness and wildlife later through his voluminous writings. Importantly, he wrote articles that were oriented toward hunters and fishermen in the sporting magazines *Outdoor Life*, *Sports Afield*, and *Field and Stream*.[14] He would have been considered an iconoclast, often articulating high standards that the agencies and others felt were too restrictive for on-the-ground managers to implement. Thus, when asked during the FDR administration to join the Fish and Wildlife Service, he declined because he didn't want to be restricted from writing his views by agency administrators.

## Jay Norwood Darling

Jay Norwood Darling (1876–1962) began his cartooning career in 1900, joining the *Des Moines Register* in 1906.[15] He signed his cartoons as "Ding" and became a political cartoonist syndicated in 130 daily newspapers. His cartoons often had wildlife subjects, and he was well-known for his support of wildlife, especially waterfowl. He was awarded Pulitzer Prizes in 1923 and 1942, and in 1934 was named best cartoonist by the country's leading editors. Darling was called "the best friend ducks ever had" and told land developers that "ducks can't lay eggs on picket fences."

In July 1934, President Franklin Roosevelt asked Darling to head the US Biological Survey, the forerunner of the US Fish and Wildlife Service. He accepted with the provision that he would return to the *Des Moines Register* in eighteen months. As chief of the US Biological Survey, Darling battled for greater national attention and expenditures for conservation. He established the Migratory Bird Conservation Commission and made great strides toward bringing hunters and other conservationists together. He also pioneered leadership in the field of professional wildlife management and initiated the Federal Duck Stamp Program, which uses proceeds from the sale of duck hunting stamps to purchase wetlands for waterfowl habitat.

Darling urged President Truman to create a wildlife refuge on Sanibel Island, Florida, after developers sought to purchase an especially valuable section of wildlife habitat on the island. Truman signed the executive order creating the Sanibel National Wildlife Refuge in 1945, and it was renamed the J. N. "Ding" Darling National Wildlife Refuge in 1967.

## Contributors to Policies and Legislation in the United States

The American Game Policy of 1930 provided impetus for subsequent legislation and wildlife management practices that underpin much of wildlife management in North America today. It was created by fourteen individuals, including Congressman Willis Robertson, and was chaired by Aldo Leopold. Much of its content is included in Leopold's book *Game Management* (1933). The 1930 policy stated clearly that the general public should bear the costs of managing and conserving wildlife.

This policy was updated in 1973 as changing agricultural and forestry practices, increasing urbanization, and changing human attitudes warranted a reexamination. Durward Allen, a faculty member at Purdue University who was active on the national level in wildlife matters and author of *Our Wildlife Legacy*,[16] chaired the committee. This committee of twenty individuals included Enrique Beltran of Mexico and Eugene F. Bossenmaier of Manitoba, plus state and federal officials across the United States. There were thirteen honorary members, including three from Ontario, Canada. The Wildlife Management Institute was the major supporter of this effort, primarily through its president, Daniel A. Poole. Poole also served as secretary of the committee. Both the 1930 and 1973 policies discussed inducements for landowners to practice wildlife habitat management and allow access for hunting. The 1973 policy specifically stated that society should protect the right of people to hunt and otherwise engage the wildlife resource.

Important legislation that stemmed, in part, from recommendations in the 1930 Game Policy helped ensure retention of North American Wildlife Model practices in the United States. The Pittman-Robertson (PR) Act, formally designated the Federal Aid to Wildlife Restoration, was passed in 1937. Senator Key Pittman of Nevada, chair of a committee on conservation of wildlife resources, and Congressman Willis Robertson of Virginia, chair of a similar committee in the House, introduced the legislation that was passed swiftly in both houses of Congress without the usual hearings and debate. The law provided for an 11 percent excise tax on sporting firearms and ammunition and 12 percent on handguns, bows, and arrows. It required states to specify that hunting licenses sales would be directed solely to the fish and wildlife agencies and not diverted to other uses; they were also to provide 25 percent of the revenues involved while the federal government would match the remaining 75 percent. Allocation of funds was based on a state's population and the amount of land, so no state would receive more than 5 percent nor less than 0.5 percent of the available revenue. This law virtually assured that all hunters, regardless of socioeconomic status, contributed directly to supporting wildlife resources, an important underpinning of the North American Model.

Senator Key Pittman (1872–1940) emerged as a major player in developing effective wildlife legislation over his long political career (1913–1940).

Educated as a lawyer, he practiced law in Seattle, then participated in the Klondike Gold Rush from 1897 to 1901. He settled in the mining town of Tonopah, Nevada, where he practiced law and after one failed try, was elected senator. He became president pro tempore of the US Senate and chaired the Senate Committee on Foreign Relations. The Key Pittman Wildlife Management Area north of Las Vegas, Nevada, was named in his honor and is today a major waterfowl hunting area.

Willis Robertson (1887–1971) co-introduced the PR Act along with Senator Pittman. Educated at the University of Richmond in law, he served in the Virginia State Senate and chaired the state's commission on game and inland fisheries from 1926 to 1932. He was elected to the US House of Representatives in 1932, served until 1946, then moved to the US Senate in 1946, and served through 1966. His experience with the Virginia games and fisheries commission made him a logical person to advance the federal aid to wildlife legislation. Other important contributors to the PR Act included J. N. "Ding" Darling, Ira N. Gabrielson, his successor as chief of the Biological Survey, Thomas Beck of the Wildlife Management Institute, and Aldo Leopold. Pittman was a democrat, while Robertson, Darling, and Leopold were conservatives; together, they exemplified the bipartisan political support the PR enjoyed.

## Contributors to Policies and Legislation in Canada

In the early decades of the last century, wildlife conservation in Canada received a great deal of attention, much as it did in the United States. Fish and wildlife resources were at all-time lows, and excessive exploitation was rampant. Many of the initiatives in Canada were accomplished under the auspices of Sir Wilfrid Laurier (1841–1919), prime minister of Canada from 1896 to 1911 and the major political figure of his time.[17] Born in Quebec and educated in law at McGill University, he became a highly skilled politician who sought out compromises on difficult issues, including promotion of national unity when cultural conflicts were severe. He directed policy through his cabinet appointments and positively affected wildlife conservation. His concern over the destruction of the forests, in particular, was evident as early as 1906. The Commission on Conservation was established in 1909 to provide scientific advice on natural and human resources and was intended to encourage better forest management practices and protection of wildlife. Laurier was successful in getting a second transcontinental railway constructed as a means of developing the western regions of Canada and was also responsible for adding Alberta and Saskatchewan to the dominion in 1905.

Sir Clifford Sifton (1861–1929) was elected to Canadian Parliament in 1896 and appointed federal minister of the interior and superintendent general of Indian Affairs by Laurier shortly thereafter.[18] He graduated from Victoria College with a degree in law and moved to Manitoba where he was admitted to the Manitoba Bar in 1882. He became involved in Manitoba politics and was attorney general for the province during the development of the national school system that Canada uses today. While he is best known for promoting immigration to western Canada, he chaired the Commission on Conservation from 1909 to 1918. This commission undertook the first large-scale survey of Canadian natural resources and associated issues. "Wise management" was an underlying premise directing commission issues and underpins the North American Conservation Model. This commission was instrumental in developing the National Parks Act, the Northwest Territories Game Act, and the Migratory Birds Convention Act.[19] Sifton was also responsible for administering the Yukon during the Klondike gold rush.

C. Gordon Hewitt (1885–1920) was educated in zoology at Victoria University of Manchester, graduating with a doctorate in science in 1909.[20] He lectured in zoology at the university until he was selected to become dominion entomologist of Canada. He was instrumental in the development of preserves and sanctuaries for wildlife and was appointed secretary of an interdepartmental advisory board on wildlife protection in 1909. He was involved in drafting the Migratory Bird Treaty in 1916 and in protecting northern wildlife through the Northwest Game Act of 1917.

Percy Taverner (1875–1947) was a self-taught naturalist with an interest in birds.[21] He was educated at Port Huron and Ann Arbor, Michigan, as an architect, but this did not deter his increasing interest in birds. In 1911, he was appointed ornithologist at the National Museum of Canada, where he served until 1942, and played an important role in establishing Point Pelee National Park in 1918 and the bird sanctuaries of Bonaventure Island and Perce Rock in 1919.[22] His writings included *Birds of Canada* (1934), which helped make bird-watching popular. He was president of the Ottawa Field Naturalists' Club and founded the *Canadian Field-Naturalist* journal. His lack of formal training notwithstanding, he was a major influence on Canadian ornithology and conservation.

The British North American Act of 1867 specifically granted the provinces the right to manage natural resources, including fish and wildlife. Seasons and bag limits were established by the provinces in these early years, often as part of legislation that addressed logging and mining as well. Andrew Blair (1844–1907), who was a provincial attorney who was New Brunswick's premier from 1883 to 1896, was instrumental in these efforts.[23] He graduated from Fredericton Collegiate School, was admitted to the New Brunswick Bar in 1866, and was elected to the House of Assembly in 1878. As premier, his Crown lands policies allowed the lumber interests to acquire leases that were either renewable or good for twenty-five years, in order to encourage planning and movement of timber. Employment of full-time forest rangers was initiated in 1892. A game act in 1893 consolidated earlier legislation, provided for game production, and revised a system for leasing salmon rivers. Wild lands were recognized as having value for attracting tourists as well. Blair was elected to the House of Commons in 1896 and was sworn in as minister of railways and canals during the Laurier years. His time as premier of New Brunswick was when he was most influential in natural resource matters. Subsequently, New Brunswick initiated registration of hunting and fishing guides in 1906, which the other provinces subsequently followed.[24]

Hamilton Mack Laing (1883–1982) was born in Ontario but is best known as a major naturalist who advocated protection of the Comox Valley of Vancouver Island.[25] The valley is an important area of global significance for birds. Laing collected specimens for the National Museum of Canada. He wrote in most of the outdoor nature magazines and was a life-long hunter. Laing's philosophy included active management to benefit conservation, and he was representative of the dichotomy in many of the earlier naturalists in advocating both predator control and protection of nature. He was an active contemporary of many ornithologists in Canada and the United States.

James Harkin (1875–1955) was born in Ontario and worked as a newspaperman from 1893 to 1901.[26] After serving as secretary to two ministers of the interior, Clifford Sifton and Frank Oliver, from 1896 to 1911, he was appointed the first commissioner of the Canadian National Parks Branch in 1911. Considered an excellent political analyst and communicator, he followed in the footsteps and philosophies of John Muir and Henry David Thoreau and was committed to preserving the recreational, aesthetic, and spiritual values of the unspoiled areas within the parks to ensure that citizens would be able to appreciate the finest scenery in the country. Harkin was also extensively involved in the approval of the Migratory Bird Treaty by Canada in 1916. The treaty was empowered by the Migratory Bird Convention Act in 1917, when a migratory birds section was established in the Parks Branch of the Canadian government.[27] Subsequently the provinces modified their laws to conform with the treaty and the act. Harkin created Wood Buffalo National Park and was instrumental in protecting pronghorn antelope in

Canada by establishing a park in Alberta. He oversaw creation of Elk Island, Mount Revelstoke, Point Pelee, Kootenay, Prince Albert, Riding Mountain, Georgian Bay Islands, and Cape Breton Highlands national parks.

Hoyes Lloyd (1888–1978) was born in Ontario and developed an interest in birds as a young boy. He was educated in chemistry at the University of Toronto, graduating in 1910 and with an MA in 1911.[28] He became associated with the American Ornithologist Union (AOU) in 1916 and was president from 1945 to 1948. In his younger years, he collected birds for specimens that are now in the Royal Ontario Museum. When the Migratory Bird Treaty was signed in 1916 and the enabling act in Canada was passed in 1917, a position in the Dominion Parks Branch for an ornithologist was created, and Lloyd was selected. He subsequently was named superintendent of Wildlife Protection, which was located within the Wildlife Division of the Parks Branch at that time. This position was responsible for administering the Migratory Birds Convention Act and the Northwest Game Act, which entailed the appointment of migratory bird officers to enforce these laws.[29] Before retiring from the Canadian Wildlife Service in 1943, he was responsible for hiring many wildlife biologists and migratory bird officers who are well represented in the professional literature. He remained active in the AOU after retirement and attended numerous meetings of several notable organizations, most notably the Federal-Provincial Wildlife Conference, the American Game Conference, and the North American Wildlife and Natural Resources Conference. He also continued to engage at international conferences on various wildlife issues throughout his life. He was an honorary member of The Wildlife Society and received the coveted Leopold Award in 1956.

C. H. D. Clarke (1909–1981) conducted major studies of muskoxen in the Thelon River drainage in the Northwest Territories in the 1930s, and these were the beginnings of efforts to manage and protect northern Canadian wildlife.[30] Born in Ontario and educated at the University of Toronto, where he studied ruffed grouse for a dissertation, he was hired by R. M. Anderson at the National Museum of Canada to do surveys on the Lake Superior shore and then to investigate muskox in the Thelon River area. This work led to the establishment of the Thelon Sanctuary. Clarke also held posts in the Parks and Wildlife Section of the Canadian federal government, and then in 1942 returned to the Mackenzie Delta region, where he recommended increased Indigenous Peoples' involvement in management of the reindeer herds. Efforts to include Indigenous Canadians in wildlife management projects have been a major endeavor ever since. In 1943 and 1944, Clarke conducted a faunal survey along the Alaskan Highway and helped establish the Kluane Reserve in the southern Yukon. He moved to the Department of Lands and Forests in Ontario in 1944, eventually becoming chief of the Division of Fish and Wildlife. He received the Leopold Medal from The Wildlife Society and was elected president of The Wildlife Society and the Canadian Wildlife Federation. Like many of the early pioneers in conservation who undertook work in areas where humans were few and the land was untouched, he was reluctant to revisit the places he once studied for fear that they would have changed for the worse.

Ian McTaggart Cowan (1910–2010) is considered a father of Canadian conservation.[31] Born in Scotland, he developed an early interest in rabbits and other small mammals, and his family had a strong tradition of hunting and fishing. In 1913 he moved to Vancouver, British Columbia, where his interests in birds and mammals grew. He received a PhD from the University of California in 1935 and was hired as a biologist for the Provincial Museum in Victoria, undertaking a number of trips across the province collecting birds and mammals for the museum. He then joined the faculty of the University of British Columbia (UBC) in 1940 and remained there until 1975. He founded the first university wildlife department in Canada and was the first to develop nature programs for television. Upon retiring from UBC, he became chancellor of the University of Victoria, where he served from 1979 to 1985. Cowan was a major influence on wildlife conservation and education in North America, receiving awards and recognition from both Canada and the United States. His students came from across Canada, the United States, and Europe. Many of them subsequently became major contributors to wildlife conservation in their own right and furthered Cowan's enduring legacy.

Many of the early wildlife conservationists in North America were well known to each other across the international border. A society of conservationists called the Brotherhood of the Venery (referred to as "B") was initiated in 1925 to advance wildlife knowledge and protection and spread the ideals of sportsmanship.[32] Hoyes Lloyd and James Harkin were the initial instigators, but Jim Munro, Percy Taverner, Kenneth Racey, Hamilton Mack Laing, Ian McTaggart Cowan, Aldo Leopold, T. Gilbert Pearson, John Burnham, Herbert L. Stoddard, and George Bird Grinnell were also members. These were the prime movers for protecting the remnant bison, regulating hunting, and creating wildlife preserves, national parks, and other sanctuaries for wildlife. The brotherhood advocated for the North American Wildlife Conservation Model through public access to hunting areas and other wildlife-oriented areas. The society gave members who were wildlife agency employees a means of objecting to policies and decisions by administrators. Through its many well-known members, this society played an important role in shaping legislation and various other aspects of wildlife conservation during its time.

## Conclusion

While there were earlier efforts espousing recreational hunting, the establishment of reserves for wildlife, and, in general, a sustainable-use approach to natural resource conservation, these issues appear to have come to the fore with the political leadership of Sir Wilfrid Laurier in Canada and Theodore Roosevelt in the United States. Some today emphasize a dichotomy between the protection of lands and wildlife and active wildlife management, but Laurier and Roosevelt, along with many of their influential contemporaries, advocated for both. This may be best exemplified by the actions of Roosevelt, who used both John Muir and Gifford Pinchot as advisors, established wildlife refuges and parks as well as national forests, and advocated strongly for hunting as an incentivizing conservation mechanism. Wilfrid Laurier and Clifford Sifton are the Canadian counterparts to this example.

In both Canada and the United States, politicians and administrators from both the conservative and the liberal sides were involved in wildlife conservation in the early years. Laurier, himself liberal, appointed conservatives such as Andrew Blair. Franklin Roosevelt appointed Ding Darling, a conservative, to his administration. Leopold, a conservative, advised Roosevelt and others regardless of whether they were liberal or conservative. These leaders were endeavoring to get qualified people into their administrations and sought advice from individuals regardless of philosophical differences.

The North American Conservation Model was not a part of the conversations of early conservation activists but has become a common description of our wildlife management and conservation practices in recent time. However, its components were conceived, understood, and promoted in the early years, perhaps best exemplified in legislative circles by the Federal Aid to Wildlife Restoration of 1937 and the Migratory Bird Treaty of 1916.

In the formative days of wildlife conservation, the active community was smaller and the people tended to be familiar with each other, on both sides of the international border. Leopold, for instance, knew Ian McTaggart Cowan and Herbert Stoddard and had correspondence with many of the people mentioned above. Virtually all of these early conservationists started out early in life with an interest in nature, more often in the wild birds they could readily observe, rather than in mammals. Their outdoor experiences frequently included either making collections of specimens for museums or trapping for fur and hunting for food and recreation. While some were certainly interested in setting aside land specifically for wildlife habitat and conservation, more often lands were set aside to control excessive logging and overgrazing, with other resources, such as wildlife, assumed to benefit secondarily. Then, too, market hunting was reducing wildlife to levels nearing extinction and ultimately became a major impetus for conservation.

While much has changed since the days when the first principles of the North American Model were being developed, what has endured is an approach to wildlife conservation that is broadly shared between Canada and the United States. Also unchanged is the ongoing close relationship between sustainable use of wildlife and the conservation mechanisms operating across their borders. What is further remarked in the careers of the conservation activists and visionaries presented here is the inclusive value for both protection and use of wildlife that most, if not all of them, espoused. The future of conservation in North America and of the Model itself will surely require such tolerance and broad vision.

*NOTES*

1. G. B. Grinnell, "American Game Protection," in *Hunting and Conservation*, ed. G. B. Grinnell and C. Sheldon (New Haven, CT: Yale University Press, 1925), 201–257.
2. D. Brinkley, *The Wilderness Warrior: Theodore Roosevelt and the Crusade for America* (New York: Harper Collins, 2009).
3. C. E. Semcer and J. Posewitz, "The Wilderness Hunter: 400 Years of Evolution," *International Journal of Environmental Studies* 70, no. 3 (2013): 438–447.
4. J. B. Trefethen, *An American Crusade for Wildlife* (New York: Winchester Press, 1975).
5. United States Department of Agriculture, "Our History, United States Forest Service," May 2014, https://www.fs.fed.us/learn/our-history.
6. Brinkley, *The Wilderness Warrior.*
7. Trefethen, *An American Crusade.*
8. Trefethen, *An American Crusade.*
9. D. Brinkley, *The Quiet World* (New York: Harper Collins, 2011).
10. Trefethen, *An American Crusade.*
11. S. L. Flader, *Thinking like a Mountain* (Lincoln: University of Nebraska Press, 1974); C. Meine, *Aldo Leopold: His Life and Work* (Madison: University of Wisconsin Press, 1988).
12. Aldo Leopold, *Game Management* (New York: C. Scribner's Sons, 1933).
13. A. H. Carhart, *Hunting North American Deer* (New York: MacMillan, 1946).
14. T. Wolfe, *Arthur Carhart Wilderness Prophet* (Boulder: University Press of Colorado, 2008).
15. Trefethen, *An American Crusade.*
16. D. L. Allen, *Our Wildlife Legacy* (New York: Funk and Wagnalls, 1962).
17. *The Canadian Encyclopedia*, s.v. "Sir Wilfred Laurier," http://www.thecanadianencyclopedia.ca/en/article/sir-wilfred-laurier/.
18. *The Canadian Encyclopedia*, s.v. "Sir Clifford Sifton," http://www.thecanadianencyclopedia.ca/en/article/sir-clifford-sifton/.
19. J. A. Burnett, *A Passion for Wildlife* (Vancouver: University of British Columbia Press, 2003).
20. *Dictionary of Canadian Biography*, s.v. "Hewitt, Charles Gordon," http://www.biographi.ca/en/bio/hewitt_charles_gordon_14E.html.
21. J. L. Cranmer-Byng, "A Life with Birds: Percy A. Taverner, Canadian Ornithologist, 1875–1947," *Canadian Field-Naturalist* 110 (1996): 1–254.
22. Burnett, *A Passion for Wildlife.*
23. *Dictionary of Canadian Biography*, s.v. "Hewitt, Charles Gordon," http://www.biographi.ca/en/bio/blair_andrew_george_13E.html.
24. Burnett, *A Passion for Wildlife.*
25. R. Fisher, "Review of Richard Mackie: 1985. Hamilton Mack Laing: Hunter-Naturalist," *BC Studies: The British Columbia Quarterly*, no. 73 (1987): 67–69.
26. *The Canadian Encyclopedia*, s.v. "James Bernard Harkin," http://www.thecanadianenclyclopedia.ca/en/article/james-bernard-harkin/.
27. Burnett, *A Passion for Wildlife.*
28. C. H. D. Clarke, "In Memorium: Hoyes, Lloyd," *The Auk* 96 (1979): 402–406.
29. Burnett, *A Passion for Wildlife.*
30. C. Norment, "C. H. D. Clarke (1908–1981)," *Arctic* 41 (1988): 256–257.
31. J. Lavoie, "Saanich Ecologist 'Last Renaissance Man,'" *Times Colonist*, April 2010.
32. B. Penn, *The Real Thing: The Natural History of Ian McTaggert Cowan* (Victoria, BC: Rocky Mountain Books, 2015).

James L. Cummins

# Critical Legislative and Institutional Underpinnings of the North American Model

*It was the rigorous advocacy of individuals and organized public efforts including sportsmen's clubs, conservation organizations, and scientific agencies that galvanized the conservation movement in North America and ultimately changed public attitudes and perceptions of wildlife and natural resource use. This progression necessitated the development of legal frameworks for conservation that would exceed and survive political change. This chapter explores the legal mechanisms underpinning the North American Model, such as the Public Trust Doctrine, the Migratory Bird Convention and Treaty, and the Lacey Act. It also highlights innovative funding mechanisms implemented through legislation to support conservation of wildlife and habitat.*

## Introduction

The development of wildlife conservation efforts has a long history and can be traced to early civilization, when many such efforts focused on hunted species. One of the earliest written policies was recorded by Marco Polo, who noted in his diaries that Kublai Khan (circa 1259–1294 AD) had ordered that there be no hunting wildlife from March through October, the breeding period, and that, for preferred species, food plots would be established, a winter feeding program would be maintained, and landscape cover would be managed.[1] Such prescriptions, dealing with hunting and land or population management, have many modern counterparts.

Indeed, equitable access to wildlife and hunting land is a key foundation for one of the longest standing and most effective approaches to conservation, the North American Model of Wildlife Conservation. Equitable access in this North American context refers to the right of a person, even if not a landowner, and irrespective of economic status, to access land and wildlife to hunt.[2]

## Influence of English Game Law

Since the Norman Conquest, English law predicated that ownership of wildlife and authority to grant hunting rights resided with the king. For the next eight hundred years, the monarchy jostled with the aristocracy for control, while both used this legislation to oppress commoners. Not only was hunting access limited, but the actual ownership of hunting equipment, dogs, and guns was reserved for those "qualified" by property ownership to hunt.[3] The original English game acts restricted hunting, and thus possession of and expertise with firearms, to the king and higher nobility. Later game acts restricted hunting to major landowners.[4] This use of hunting rights to control the populace and ownership of firearms was likely in the minds of framers of the United States' Constitution when drafting the Bill of Rights.

The most dramatic change in the English game acts came after the English Civil War of the 1650s. Before the war, lands were held by grant from King Charles I, who had the right to hunt anywhere and had also set aside all the royal forests for his private use.[5] Charles I was beheaded, but eventually his son, Charles II, was recalled from exile to become king, and, with the gentry's return to power, instituted the Game Act of 1671. This 328-year-old statute and subsequent regulations passed by the aristocracy superseded earlier laws without repealing them and created harsh provisions, such as a "gentleman's game privilege," which limited hunting rights to those owning vast amounts of property. This legislation also empowered such landowners and their appointees to search the homes of those designated not qualified to hunt and to seize dogs, guns, and other hunting equipment.[6] Meanwhile, those authorized as hunters could hunt not only on their own land but also on another person's private property as well.[7]

After Charles II's death, his brother James II came to power and also used the Game Act to aggressively disarm the English citizenry. As he became more and more unpopular and insecure, he ordered mass searches and the confiscation of any guns from owners not authorized to possess weapons.[8]

In 1688, James II was overthrown. Parliament composed a Declaration of Rights, which the new rulers were required to uphold. The declaration protected several rights, including the right of Protestants to have weapons for defense. (Although Catholics were a minority, Parliament rarely disarmed them.) Soon afterward, Parliament amended the Game Act to remove its prohibition on weapons; however, the restriction of hunting to the wealthy landowners remained the law.[9]

## Colonial and United States Game Law

In early European America, colonists and their governments faced different social and environmental circumstances. English, French, Spanish, and Dutch colonies attacked each other, Indians raided all of them, and dangerous wildlife such as bears, mountain lions, and wolves were common. In response, the early colonies encouraged private gun ownership and provided liberal hunting rights, circumstances that would eventually require policies to regulate wildlife takings and conserve populations.[10]

Measures to bolster the population of white-tailed deer, a staple in the diet of colonial America, were undertaken as early as the 1630s in the form of bounties on predators. An early Virginia law allowed hunting of everything except wild hogs—but even there the law provided that the bounty for killing a wolf would be a license to kill a wild hog. By 1646, with white-tailed deer numbers not responding, Portsmouth, Rhode Island, enacted America's first closed season on deer, with no hunting allowed between May 1 and November 1 and a £5 fine for any violation. In 1699, the Colony of Virginia also established a closed season, with no hunting from February 1 through July 31. The fine for violation of this law was five hundred pounds of tobacco, an incredible sum at the time. In 1705, in an effort to improve compliance with the law, an amendment was made providing for the payment of half the fine to anyone acting as an informant.[11]

The development, maintenance, and enforcement of many early wildlife management policies worked better on paper than in practice. Because venison and white-tailed deer hides had become staples of everyday colonial life, nearly every law passed to protect deer in Virginia also contained an exemption for settlers living on the western frontier. Meanwhile, laws continued to evolve, and by 1738 separate hunting seasons had been established for bucks versus does and fawns, and the fines for poaching were converted from tobacco to twenty shillings. There was an added stipulation that if a fine was not paid within six months, the offender would receive twenty lashes.[12]

Thus, in the New American colonies, hunting and gun ownership were seen as essential rights, not as components of royal prerogative or privilege. All these rights were amplified by the American

Revolution (1775–1783), in which the citizen soldier, already trained in the use of his personal firearm, successfully fought and defeated the best army in the world.[13] The relationship between personal freedoms, gun ownership, and hunting was greatly strengthened in the drive for American independence.

Following the defeat of Britain, the American people eventually settled on a constitution in 1787. However, it lacked a bill of rights. Many powerful people at the time criticized the Constitution because it gave powers to the government, but contained no rights to protect the people.[14] Several states demanded a bill of rights as a condition of ratifying the original Constitution. The Bill of Rights would, of course, include the Second Amendment, which guaranteed the right to keep and bear arms.[15] Not surprisingly, Pennsylvania also called for a right to hunt and fish in the Bill of Rights,[16] and, in 1777, the State of Vermont included a constitutional right to hunt as a provision in its state constitution.[17] In 1789, the first Congress proposed and debated what eventually became the United States Bill of Rights. Rather than Norman common law, its basis was earlier Germanic common law, which held that "freemen" possessed allodial rights, by which they were free to bear arms, profit from unclaimed resources, and hold land without obligation to a sovereign.[18] These rights were to have considerable influence on the trajectory of resource use and the wider conservation movement in the United States.

## Exploitation versus Appreciation

Hunting as a means of survival was eventually accompanied by increasingly heavy commercial demands on wildlife populations and on white-tailed deer in particular. In addition to meat provisioning, a significant trade in deer hides also developed. Records indicate that between 1698 and 1715 Virginia annually exported approximately 14,000 deer hides to Europe. Such harvests would quickly escalate, however, and in a single year (1748), 160,000 deer hides were sent to England from South Carolina. This industry was limited to the eastern United States. Additionally, market hunting for deer to feed a growing nation became widespread. This harvest reached a peak in the mid- to late 1800s. Accounts from this period include that of an individual market hunter from northwestern Virginia who reportedly killed over 2,500 deer prior to 1860. A father and son in Minnesota were said to have killed 6,000 white-tailed deer for the market in a single year (1860).[19] These are but a few examples of the extensive wildlife exploitation occurring in the United States at that time. Further market hunting on the Western frontier was actually promoted by the United States military. In light of the utter failure of George Custer's 1867 campaign against Native Americans, Generals William T. Sherman and Philip H. Sheridan developed an unofficial plan to deal with the "Indian problem" through destruction of the wildlife the various tribes relied upon for food, clothing, and other provisions. By protecting market hunters and their transit via the recently built railroad, the military further exacerbated commercial hunting and the decimation of wildlife populations and the Native peoples who depended upon them.[20]

While hunting was undoubtedly required for survival in the colonies, people were, at the same time, interested in the scientific aspects of wildlife. The study of living natural resources, especially the surveying of them, emerged early in North America. About the time the United States was founded in 1776, William Bartram, an American naturalist from Philadelphia, Pennsylvania, traveled throughout the eastern United States to describe its flora and fauna.[21] Such efforts continued, and in the early 1800s, naturalist John James Audubon (1785–1851) spent decades studying and drawing North American bird specimens. Audubon was born in what is now known as Haiti, but was reared by his stepmother in France. At the age of eighteen, he was sent to America, in part to escape being drafted into Napoleon's army. He lived on the family-owned estate near Philadelphia, where he hunted, studied, and drew birds. Though Audubon was not the first person to attempt to paint and describe the many birds of America, he certainly was among the most gifted and influential. Like many citizens of his era, he saw a natural linkage between hunting, the story of natural history, and a wider appreciation of the natural world.

## Early American Wildlife Laws

As market hunting and overexploitation devastated North America's natural resources, it became clear to many that a philosophical shift was essential in order to preserve a natural inheritance for future generations. Ancient Roman law categorized wildlife, and nature as a whole, as common property that could not be privately owned[22] and a revised American philosophy acknowledged this, while preserving the allodial rights of property and arms ownership cemented under the Bill of Rights. Further, while the American Revolution had eliminated the monarchy as trustee of communal property, including wildlife, in 1842 the Supreme Court recognized individual states as public trustees (*Martin v. Waddell*) and in 1896 designated wildlife as a public trust resource (*Geer v. Connecticut*).[23] Subsequent legislation retained wildlife as a public trust, and the concept became a vital part of North American conservation policy and legislation.[24]

One of the first national conservation efforts by a federal government in North America occurred in 1871, when the US Congress created the US Commission on Fish and Fisheries under the authority of the Department of Commerce. The commission was charged with investigating the declining numbers of coastal and lake food fishes and recommending remedial measures to Congress. During the same period, the first park in North America was created when the US Congress established Yellowstone National Park in 1872, although the park's administrators lacked the legal authority and personnel to protect its resources. Ironically, it was General Sheridan who in 1886, a decade after calling for eradication of bison in Texas, sent the US Cavalry to stop the pillaging of Yellowstone and decimation of its wildlife populations by poachers and market hunters.[25]

Communicating the need for conservation, as well as best practices in hunting and fishing, was significantly advanced in 1873 with the founding of *Forest and Stream* magazine. A decade later, in 1883, Charles Sprague Sargent published a *Report on the Forests of North America, Exclusive of Mexico*, warning of the need to reform destructive timber-management policies and significantly broadening the scope of early conservation reform.

By the mid-1880s, key leaders of the American conservation movement were still emphasizing the loss of wildlife and habitats, and they called on people to fundamentally change America's approach to wild resource use. These early conservation leaders effectively utilized the press to communicate the need for conservation. In 1887, Theodore Roosevelt invited George Bird Grinnell, the editor of *Forest and Stream*, to a dinner party at his home in New York City. The guest list included distinguished writers, scientists, explorers, military leaders, industrialists and political figures. The one thing they shared was a passion for big game hunting and traveling in the American West. At this gathering, Roosevelt discussed his idea to halt the destruction of America's wildlife. Hence, the Boone and Crockett Club was born, articulating a sustainable and wise-use philosophy to address the decline and recovery of North America's natural resources. The members pledged to promote rifle hunting as a sport, to conserve big game populations and their habitats, and to promote conservation legislation. The Boone and Crockett Club's goals were promoted through their publications, which Grinnell and Roosevelt edited, and through Grinnell's commentaries in *Forest and Stream*. In addition to efforts within the United States, a group of influential Canadians shared the Boone and Crockett Club's goals of restoring wildlife resources. Sir Clifford Sifton, Dr. Charles Gordon Hewitt, and Dr. Cecil C. Jones, to name a few, shared friendship, information, and influence to establish similar systems of wildlife conservation in the United States and Canada.[26]

These early conservation leaders initially focused on protecting wild places and impeding the killing of game and fish for markets. The result of the Boone and Crockett Club and their Canadian counterparts' efforts to establish a foundation and framework for conservation in America were to strongly influence what has become known as the North American Model for Wildlife Conservation.

With the early successes of the United States' first conservationists, many felt the need to expand on those efforts. From General Sheridan's aforementioned change of heart sprang thirty-two years of military oversight that rescued the idea of national parks, as well as prevented the extinction of plains bison.[27] Formation of the forest reserves and their evolution into national forests, development of treaties to conserve migratory birds, passage of the Pittman-Robertson Act, and other movements came to fruition because people agreed on the need for conservation and worked together to implement change. History shows that unified efforts do produce results for conservation of fish and wildlife and their habitats.

These conservationists played a major role in impeding the massive killing of wild animals for meat, hide, and plume markets, which resulted in the development and support of the Lacey Act and other modern day game laws. Other significant pieces of legislation included the Reclamation Act, National Wildlife Refuge System Act, Migratory Bird Conservation Act, Federal Aid in Wildlife Restoration Act, Healthy Forests Restoration Act, and what is commonly called the "Farm Bill."

While many early conservation efforts focused on hunting and natural resource use, strenuous efforts focusing on protection of landscapes also emerged. In 1892, John Muir founded the Sierra Club, an environmental organization whose mission is to explore, enjoy, and protect the wild places of the earth; to practice and promote the responsible use of the earth's ecosystems and resources; to educate and enlist humanity to protect and restore the quality of the natural and human environment; and to use all lawful means to carry out these objectives. This philosophy emphasizes protection more than sustainable resource use. Conservation can be defined as the careful utilization of a natural resource to prevent depletion, injury, decay, waste, or loss, which would seem to embrace an inclusive perspective.[28] Over time, however, conflicts between protectionist and utilitarian views have emerged, and these are conveyed in various laws pertaining to wildlife and habitats.

The *World English Dictionary* defines a conservationist as "a person who advocates or strongly promotes careful management of natural resources and of the environment." Many people today may not identify with this definition or may take for granted the efforts that laid the foundation for the conservation systems that exist today in North America. While chapter 4 provides a wide review of individuals who contributed, this chapter focuses on those who championed iconic conservation legislation, including some whose names are embedded in the laws they helped to forge.

John Fletcher Lacey (1841–1913) was an eight-term congressman from Iowa. Today, Lacey is known mostly for his namesake, the Lacey Act of 1900. This act protects both wild plants and wild animals by creating civil and criminal penalties for a wide array of violations. Most notably, the act prohibits trade in wildlife, fish, and plants that have been illegally taken, possessed, transported, or sold.[29] It is one of the most significant pieces of legislation to curtail the unlawful taking of wildlife in the United States and reflects the commitment of its originator.

Lacey once commented, "For more than three hundred years destruction was called 'improvement' and it has only in recent years come to the attention of the people generally that the American people were like spendthrift heirs wasting their inheritance."[30] The Lacey Act assisted, in a most dramatic way, to address this problem.

George Bird Grinnell (1849–1938) was an American anthropologist, historian, naturalist, and writer. He was prominent in movements to preserve wildlife and conservation in the American West. For many years he published articles and lobbied for congressional support for the endangered American bison. In 1882, Grinnell promoted an expansion of Yellowstone National Park, along with a variety of new protection measures. Although the expansion proposal was defeated, Congress did pass an amendment allowing use of troops to enforce park regulations, which enabled General Sheridan's efforts in 1886.[31] In 1894, Grinnell used a poaching incident to garner public support and eventual passage of the Yellowstone Game Protection Act, establishing laws protecting Yellowstone's wildlife and natural wonders.[32] Grinnell was editor of *Forest and Stream* magazine from 1876 to 1911. He contributed essays to many magazines and professional magazines in his lifetime and was a powerful force behind many legislation efforts.[33]

Theodore Roosevelt (1858–1919) traveled to the American West for the prospect of big game hunting, but by the time he arrived, the largest herds of bison were gone—obliterated by commercial use, by disease, and as part of the unofficial Native American genocide policies of the national government. Roosevelt stayed in the West and was a keen observer of the ongoing decimation of the wildlife there, as well as the loss of habitat for not only large animals, but small mammals and songbirds, as well.[34] His many later efforts in conservation policy and legislation were influenced by these experiences.

Early conservation efforts in land acquisition were aimed at the development and passage of the Forest Reserve Act (1891), which resulted in approximately 45 million acres for national forests being established in the United States.[35] The act gave authority to the president to withdraw land from public domain as "forest reserves" managed by the General Land Office of the Department of Interior to protect watersheds from erosion and flooding and to preserve the nation's timber supply from over-exploitation. After becoming president in 1901, Roosevelt used his authority to protect wildlife and public lands by creating the United States Forest Service and establishing fifty-one federal bird reservations, four national game preserves, 150 national forests, five national parks, and eighteen national monuments. During his presidency, Theodore Roosevelt protected approximately 230 million acres of public land.[36]

Roosevelt's most famous endeavor as a hunter and conservationist was during his second term as president, when he traveled to Mississippi to hunt black bear. Ever the sportsman, Roosevelt declined to take a black bear that famed tracker and guide Holt Collier had tied to a tree. This hunt is often referred to as the most famous hunt to have taken place on American soil and to have given birth to the Teddy Bear and the tenet of fair-chase hunting.[37]

Roosevelt's motivation for his many conservation achievements are reflected in an 1886 address, in which he stated: "We have fallen heirs to the most glorious heritage a people ever received, and each one must do his part if we wish to show that the nation is worthy of its good fortune."[38]

## Trade in Wildlife

The Lacey Act (16 U.S.C. §§ 3371 et seq.), previously discussed in reference to its author, John Lacey, is the United States' oldest national wildlife protection statute. Along with the many added necessary amendments, the Lacey Act is a powerful weapon in the fight against illegal trade, possession, and transportation of wildlife, fish, and plants. In its original form, the Lacey Act addressed the growing scarcity of domestic birds, the international trade in bird feathers, the introduction of harmful avian species, and the growing problem of interstate merchandising of illegally killed and transported wildlife.

In the late 1800s, the United States experienced a growing economy. This economic prosperity created fashion and novelty markets for merchandise such as hats adorned with feathers. By the turn of the century, millions of birds, many of them wading birds such as herons, were being killed for their plumage for the millinery trade. Traders made more money per ounce from feathers than they could from gold at that time. Intended as a means to enhance, rather than replace, state laws, the Lacey Act made it illegal to deliver for shipment, or ship parts or bodies of any wild animals or birds killed in violation of state law, required clear markings and labeling of all interstate shipments of wildlife, and removed federal restrictions on the states' ability to regulate the sale of wildlife within their borders by subjecting all game animals and birds entering a state to the state's laws.

Forgotten were fish, as the original Lacey Act did not apply to them. It wasn't until 1926 that Congress addressed the overfishing of smallmouth and largemouth bass. Bag limits set by the states failed to inhibit excessive catches due to the popularity of the species with anglers. Treasured for their fighting ability, black bass were hidden in barrels between layers of other fish and illegally transported across state lines,[39] which served only to further their depletion. Congress decided it was essential to step in before it was too late and the Black Bass Act (16 U.S.C. §§ 851 et seq.), using the Lacey Act as a model, was enacted to enhance state laws by regulating and prohibiting certain interstate shipments of bass. This act applied to any actions made by companies, corporations, associations, and common carriers, rather than just individuals.

The Black Bass Act was expanded in 1947 to cover all "game fish," as defined by state laws. Eventually it was suggested that the act be expanded to cover game fish taken illegally on lands under federal jurisdiction, including national parks and Native American reservations. This led to an amendment covering game fish taken, transported, purchased, or sold contrary to state law. In 1949, the Lacey Act added the prohibition of importation of wild animals or birds under inhumane or unhealthful conditions, and in 1952, legislation encompassing all fish rather than just game fish was adopted.

Amendments to the Lacey and Black Bass Acts in 1969, as well as the introduction of the nation's second version of the Endangered Species Act, were a testament to America's growing awareness of the need for conservation. Coverage was expanded to include mollusks, reptiles, amphibians, and crustaceans.

Finally, in 1981, prompted by reports of increased incidents of illegal trade in fish and wildlife, Congress revisited the Lacey and Black Bass Acts to redress deficiencies in these laws and combined them into a new statute called the Lacey Act Amendments of 1981. There is much to be told of the evolution of the Lacey Act over the last century, but presently it has come to encompass and prohibit two general types of activity. First, it prohibits the failure to mark, as well as falsify documentation for, most wildlife shipments, while providing civil and criminal penalties for such violations. Second, it prohibits trade in wildlife, fish, or plants that have been illegally taken, possessed, transported or sold.

Administered by the United States Departments of the Interior, Commerce, and Agriculture, the Lacey Act is enforced by agencies including the Fish and Wildlife Service, the National Marine Fisheries Service, and the Animal and Plant Health Inspection Service. In determining violations of the Lacey Act, authorities must decide whether the wildlife, fish, or plant in question is covered by said act. The act defines the terms "fish or wildlife" as

> any wild animal, whether alive or dead, including without limitation any wild mammal, bird, reptile, amphibian, fish, mollusk, crustacean, arthropod, coelenterate or other invertebrate, whether or not bred, hatched or born in captivity, and includes any part, product, egg or offspring thereof.

This definition covers essentially any wild animal, vertebrate or invertebrate, dead or alive, from any part of the world, as well as any part of, or product made from, such specimens. Thankfully, this broad definition precludes the need for constant interpretation.

The act also pertains to plants and refers to

> any wild member of the plant kingdom, including roots, seeds and other parts thereof (but excluding common food crops and cultivars) which is indigenous to any State and which is either (A) listed on an appendix to the Convention on International Trade in Endangered Species of Wild Fauna and Flora, or (B) listed pursuant to any State law that provides for the conservation of species threatened with extinction.

Therefore, originally the act applied only to plants native to the United States, and specifically protected by a state law. Given these definitions, importing a skeleton of a tiger or live snakes to the United States warranted an automatic felony, but importing an equally endangered plant species was not actionable under the Lacey Act. The lax plant restrictions were addressed when the Lacey Act was amended in 2008 to include an even wider variety of prohibited plants and plant products, including products such as paper and timber made from illegally logged woods. This amendment, which is recognized as the world's first ban on trade in illegally sourced wood products, has two major components: a ban on trading plants or plant products harvested in violation of the law; and a requirement to declare the scientific name, value, quantity, and country of harvest origin for some products. Violators of this law can face criminal and civil sanctions regardless of prior knowledge.

All amendments or objections have strengthened the Lacey Act, which now stands as one of the most comprehensive and far-reaching weapons in the federal arsenal to combat wildlife crime. As criminal activity increases in domestic, as well as international, wildlife trafficking, the Lacey Act serves as a critical protection for wildlife, domestically and internationally.

## State Fish and Wildlife Law

In the 1930s, it was a rare event to see white-tailed deer in the Eastern United States. Indeed, taking of the land for human purposes and overharvesting without regulation had caused severe declines in many species of wildlife. Yet today white-tailed deer are incredibly abundant across a large portion of North America. Conservation initiatives such as regulating hunting, species reintroduction programs, and habitat acquisition and improvement all contributed to this recovery. Similar recoveries have been recorded for numerous other wildlife species, including wild turkey, wood duck, pronghorn antelope, elk, black bear, bobcat, wild sheep, and mountain lion.

In many such instances, North American hunters and their membership organizations have played a major role in launching or funding conservation and recovery efforts. The Federal Aid in Wildlife Restoration Act (16 U.S.C. §§ 669–669i) is one law that resulted from these conservationists' efforts. This act, commonly referred to as the Pittman-Robertson Act, was sponsored by Senator Key Pittman of Nevada and Congressman A. Willis Robertson of Virginia and signed into law by President Franklin D. Roosevelt on September 2, 1937. It was created to provide funding for the restoration, rehabilitation, and improvement of wildlife habitat, the distribution of educational information, and wildlife management research. An amendment made in 1970 allowed for funding for the development and management of target ranges, as well as hunter training and education programs.

The Pittman-Robertson Act is funded through an 11 percent federal user-pay fee, or excise tax, on sporting arms, archery equipment, and ammunition and a 10 percent tax on handguns. Collected from manufacturers by the US Department of the Treasury, these funds are apportioned each year as grants to the states and territories by the US Department of the Interior on the basis of formulas set forth in the act. Grant funding is available only to qualified state government agencies.

To be eligible for federal funds, a state must assent to the provisions of the act and have in place laws governing the conservation of wildlife. In addition, states must have laws prohibiting the diversion of license fees paid by hunters for any purpose other than the administration of the state's fish and wildlife department. All projects aided under the act must be agreed upon by the secretary of the interior and the fish and wildlife departments of the state where the project is located.

Each state's apportionment is determined by considering the total area of the state and the number of licensed hunters in that state. As a cost-reimbursement program, the state covers the full amount of an approved project, then applies for federal reimbursement for up to 75 percent of its expenses. Therefore, 25 percent of the cost must come from a nonfederal source, such as license fees paid by hunters and other such sources.

This user-pay fee quickly generated tens of millions of dollars from the sale of sporting arms and ammunition, which were mandated to go back into state and local organizations and governments. As the money kept building, a repeal bill was drawn up to relieve sportsmen of the financial burden. However, because positive and dramatic results were seen nationwide, this user-pay fee was kept in place, and the decision was supported by hunters and their organizations. Of the funding available to the states, more than 60 percent is used to buy, develop, and maintain wildlife management areas. Over 4 million acres have been purchased outright since the program began, and nearly 40 million acres are managed for wildlife under agreements with other landowners. Along with land acquisition, better land management methods are also supported. Some examples in the United States include planting trees and shrubs in the Great Plains as cover to shelter quail, pheasants, and other wildlife during winter storms; creating wildlife watering holes in the arid Southwest; controlling fires and restoring waterfowl habitat in the South; and creating habitat shelter in the Northeast for rabbits, deer, woodcocks, and other wildlife.

Although funded wholly by firearm users and archery enthusiasts, the Pittman-Robertson Act benefits the broader citizenry, including those who do not hunt but enjoy such recreational hobbies as bird watching, painting, sketching, and nature photography, as well as other outdoor pursuits. Almost all lands purchased with money from this act are managed for wildlife production as well as other public uses such as hiking, fishing, and camping.

A companion piece to the Wildlife Restoration Act is the Sport Fish Restoration Act. It is also based on a "user-pay-user benefit" approach. First created in 1950, when the Federal Aid and Sport Fish Restoration Act was passed (16 U.S.C. §§ 777–777n, more commonly called the Dingell-Johnson Act after its congressional sponsors), the program diverts federal excise taxes on fishing tackle, motor boat fuel taxes, and import duties on tackle and boats to the states for sport fisheries conservation and development projects. These revenues are collected by the sports fishing industry, deposited in the US Department of Treasury, and allocated the following year to state wildlife and fisheries agencies for wildlife and fisheries projects, as well as fishing access projects.

In 1984, as a result of amendments to the Federal Aid and Sport Fish Restoration Act sponsored by Senator Malcolm Wallop of Wyoming and Representative John Breaux of Louisiana, program eligibility was expanded to include funding for other types of projects. Sources of revenue identified in the act included a 10 percent user fee on additional items of fishing tackle and a 3 percent fee on electric trolling motors and sonar fish finders. A fee is also paid on motor fuels used for recreational boating and import duties on fishing tackle, yachts, and pleasure craft. Sport fish restoration funds are added to traditional sources of state fishing agency funding (i.e., fishing license revenues) for new and expanded state fisheries programs. States cannot use the new monies derived from the Wallop-Breaux Amendment to substitute for traditional sources of revenue.

The federal aid division of the United States Fish and Wildlife Service has responsibility to allocate the Sport Fish Restoration account money utilizing the following formula: (1) 40 percent of each state's share is based on the state's land and water area, including coastal and Great Lakes waters in relation to the total land and water area of the United States, and (2) 60 percent of the state's share is based on the number of paying sports fishing licenses in relationship to all of the paid fishing licenses in the United States. However, no state may receive more than 5 percent of the total apportionment and no state may receive less than half of 1 percent.

Sport Fish Restoration Program funds are available only to state agencies responsible for managing the sports fishing resources of the state. Universities, private organizations, other state agencies, or county/municipal governments may cooperate with state fisheries agencies on Sport Fisheries Restoration projects that are administered by the state fisheries agencies, but the state's sport fisheries agency is responsible for setting priorities in making project proposals to the US Fish and Wildlife Service.

## International Treaties for Migratory Birds

In 1913, the Weeks-McLean Act, North America's first migratory bird legislation, was passed, declaring that all migratory and insectivorous birds were under protection of the United States federal government. The continent's first international commitment to managing migratory birds was formalized in the 1916 treaty between the United States and Great Britain (for Canada) for the protection and recognition of migratory birds as an international resource.

The Migratory Bird Treaty Act (16 U.S.C. §§ 703 et seq.) was passed in 1918, implementing the treaty between the United States and Great Britain (for Canada) for the protection of migratory birds. This act, which was a landmark in wildlife conservation legislation, made it unlawful to take, possess, buy, sell, purchase, or barter any migratory bird, including feathers, parts, nests or eggs. It also provided for the regulation of migratory bird hunting.

In 1929 Congress passed the Migratory Bird Conservation Act (16 U.S.C. §§ 715 et seq.), also known as the Norbeck-Anderson Act, which created the Migratory Bird Conservation Commission to consider and approve any areas of land and/or water recommended by the secretary of the interior for purchase or rent by the US Fish and Wildlife Service, and to fix the price at which such areas may be purchased or rented. The commission also considers the establishment of new waterfowl refuges.

Congress passed the Migratory Bird Hunting Stamp Act (16 U.S.C. §§ 718; Pub. L. 109-266) in 1934. Popularly known as the Duck Stamp Act and renamed the Electronic Duck Stamp Act of 2005, it requires the purchase of a special tag by waterfowl hunters, which supports the Migratory Bird Conservation Fund. Ninety-eight percent of the revenue from the stamp goes toward the acquisition of wetlands important to waterfowl feeding, staging, and breeding.

In 1972, the United States signed the Migratory Bird Treaty with Japan and amended the Migratory Bird Treaty with Mexico to protect additional species, including birds of prey. The governments of both the United States and Canada signed the North American Waterfowl Management Plan to conserve waterfowl and migratory birds in North America in 1986. This was in response to the loss of at least 53 percent of wetlands in the contiguous United States and a minimum of 29 percent of wetlands in Canada, which had led to plummeting populations of waterfowl.[40] The plan was expanded to include Mexico in 1994.

## The Environmental Movement

After World War II, growing concerns for the environment led to significant new legislation on a variety of fronts. While these efforts were not specifically directed toward wildlife conservation, their impact on improving and protecting environmental quality generally has obvious relevance to fish and wildlife populations. For example, in 1962, Rachael Carson published *Silent Spring*, documenting the detrimental effects of indiscriminate use of pesticides on the environment, particularly on birds. It is one of the two most significant environmental books of the twentieth century (the other being *A Sand County Almanac* by Aldo Leopold). Carson's book marked the beginning of the modern-day environmental movement.

In 1948, Congress passed the Federal Water Pollution Control Act (Pub. L. 80-845) to restore and maintain the chemical, physical, and biological integrity of the nation's waters. This was the forerunner to what we know today as the Clean Water Act (33 U.S.C. §§ 1251 et seq.). Congress passed the Water Quality Act in 1965, directing the states to develop water quality standards. A federally directed initiative was deemed necessary, since many watersheds and waterways crossed state boundaries. Congress amended the Federal Water Pollution Control Act to establish goals of eliminating releases of high amounts of toxic substances into water, eliminating additional water pollution by 1985, and ensuring that surface waters would meet standards necessary for human sports and recreation.

The Water Quality Act of 1987 (33 U.S.C. §§ 1251 et seq.) made substantial additions to the Clean Water Act and directly affected the standards program. This act provided a new approach to controlling toxic pollutants by requiring "states to identify waters that do not meet water quality standards due to the discharge of toxic substances, to adopt numerical criteria for the pollutants in such waters, and to establish effluent limitations for individual discharges to such water bodies." As amended in 1987, the Clean Water Act requires that states adopt numeric criteria for toxic pollutants listed under section 307(a) of the Clean Water Act for which section 304(a) criteria have been published, if the presence of these pollutants is likely to adversely affect the water body's use. Additionally, for the first time, the act explicitly recognizes anti-degradation, or the requirement that states maintain the quality of a body of water so as to sustain its existing uses.

Congress passed the Oil Pollution Act to mitigate and prevent civil liability from future oil spills off the coasts of the United States. It established limitations on liability for damages resulting from oil pollution and created a fund for the payment of compensation for such damages. It forms part of oil spill governance in the United States. The law states that companies must have a "plan to prevent spills that may occur" and have a "detailed containment and cleanup plan" for oil spills (33 U.S.C. §§ 2701 et seq.).

In 1955, Congress passed the Air Pollution Control Act (Pub. L. 84-159) to provide research and technical assistance relating to air pollution control. This was the first federal air pollution legislation. The act funded research on scope and sources of air pollution. Congress passed the Clean Air Act in 1963 (42 U.S.C. §§ 7401 et seq.), creating a research and regulatory program in the US Public Health Service authorizing the development of emission standards for stationary sources, but not mobile sources. In 1970, Congress greatly expanding the federal mandate by requiring comprehensive federal and state regulations for both industrial and mobile sources of pollution with increased enforcement authority. The act established National Ambient Air Quality Standards and National Emission Standards for Hazardous Air Pollutants. The 1963 act was later amended, authorizing programs for acid deposition control, authorizing controls for 189 toxic pollutants, establishing permit program requirements, expanding and modifying provisions concerning National Ambient Air Quality Standards, and expanding and modifying enforcement authority.

In 1969, Congress passed the National Environmental Policy Act (42 U.S.C. §§ 4321 et seq.) "to declare national policy which will encourage productive and enjoyable harmony between man and his environment; to promote efforts which will prevent or eliminate damage to the environment and biosphere and stimulate the health and welfare of man; to enrich the understanding of the ecological systems and natural resources important to the Nation." This act also established the Council on Environmental Quality, a division of the Executive Office of the President. It was signed into law in 1970.

In 1963, Congress passed the Wilderness Act (16 U.S.C. §§ 1131 et seq.) establishing the National Wilderness Preservation System that protects federal managed wilderness areas designated for preservation in their natural condition.

In 1966, Congress passed the Endangered Species Preservation Act (Pub. L. 89-669) to "conserve, protect, restore, and propagate certain species of native fish and wildlife." Congress expanded the legislation's reach by prohibiting the importation into the United States of species "threatened with extinction worldwide," except as specifically allowed for zoological and scientific purposes and propagation in captivity. Congress passed the modern-day Endangered Species Act (16 U.S.C. §§ 1531 et seq.) in 1973 to govern the process of identifying threatened and endangered species, provide protections for such species, and govern federal actions that could affect such species or their habitat. At about the same time, in 1972 Congress passed the Marine Mammal Protection Act (16 U.S.C. §§ 1361 et seq.), establishing a moratorium on the taking and importing of marine mammals, such as polar bears, sea otter, dugongs, walrus, manatees, whales, porpoise, seals, and sea lions.

These various laws create the vast body of environmental legislation that helps protect the landscapes and natural diversity within the United States. To an appreciable extent, they are mirrored by similar legislation in Canada. Specific wildlife-oriented legislation works in common purpose with such laws to ensure a comprehensive legislative framework for conservation in both countries.

## Cooperative Wildlife Research Units

In the 1930s, Ding Darling, a Pulitzer Prize–winning cartoonist with a passion for conservation was named the first chairman of the Iowa Fish and Game Commission. Recognizing a lack of trained wildlife managers and scientific information, the commission under Darling conceived of a plan to work with what is now Iowa State University to establish a cooperative program for research in wildlife conservation. The school, the commission, and Darling himself would each contribute funds according to plan, which was agreed upon in 1932.[41]

In 1934, Darling became director of the US Department of Agriculture's Biological Survey which is now the US Department of the Interior's Fish and Wildlife Service. Darling proposed his Cooperative Wildlife Research Unit idea to Congress while searching for support to establish a nationwide program. Congress agreed and established a system consisting of four types of partners: state, federal, university, and private cooperators.[42]

While the system originally focused on wildlife, in 1960 Congress enacted the Cooperative Units Act authorizing fisheries research to be included in the unit systems programs. From the original ten Cooperative Wildlife Research Units established after 1935, there are now units at universities in thirty-eight states. The cooperative nature of the program has been maintained. The federal government funds the salaries of the unit scientists, the university provides these scientists graduate faculty appointments along with office, laboratory, and administrative support, the state fish and wildlife agency provides operational funding, and the Wildlife Management Institute provides guidance and support.[43]

Annually, each unit's coordinating committee, composed of representatives from each of the cooperators, identifies research and training needs and approves the research agenda. In the mid-2000s, all Department of the Interior research offices were transferred to US Geological Survey, which is home today to the Cooperative Fish and Wildlife Research Units. The US Fish and Wildlife Service has remained a cooperator.[44]

Throughout the evolution of the program, the mission has held steady. Its aims are (1) to conduct research to provide science solutions to fish and wildlife management needs of the cooperators; (2) to train the next generation of wildlife and fisheries professionals through master's, doctoral, and postdoctoral investigations; and (3) to provide the training and technical assistance to cooperators to help them in applying the latest science tools and technologies for conservation. Students graduating from the Cooperative Research Unit program have a particular brand associated with them. Wildlife Agencies know the students have had the advantage of having worked with academic supervisors with knowledge of wildlife and understand how to apply science to real-world conservation problems. Currently there are over one thousand projects being conducted and over five hundred graduate students enrolled nationwide in the units.[45] The long-term contributions on this program to wildlife and fish conservation have been enormous.

## Private and Agricultural Land Conservation Practices

All too often we associate only public land with wildlife conservation in North America, but private lands are vitally important to the conservation of wildlife in the United States because they constitute 70 percent of the land ownership in the lower forty-eight states. Furthermore, 50 percent (890 million acres) of the land base in the contiguous United States is managed for agricultural purposes as cropland, pastureland, and rangeland.[46]

Many of the successes in conservation and wildlife management achieved in the United States are anchored by key pieces of agricultural legislation that have been passed in the last one hundred years. One such critical piece of legislation is the Farm Bill, which had its beginning in the 1930s, when it was known as the Agricultural Adjustment Act (Pub. L. 73-10).

This initial legislation was intended to help steer the country out the Great Depression. It addressed widespread domestic hunger, falling crop prices for farmers, and the catastrophic Dust Bowl that resulted in massive soil erosion. This act restricted agricultural production by paying farmers subsidies not to plant part of their land, to let natural vegetation grow, and to remove excess livestock. Its purpose was to reduce the surplus of agricultural commodities, and therefore effectively raise the value of crops. Periodically, the legislation has been re-authorized, often improving conservation policy while addressing commodity payments such as disaster assistance payments and crop insurance, as well as supplemental nutrition assistance programs.

The Farm Bill of recent times is a compilation of many different acts that have been passed by Congress to enhance agricultural productivity and conservation on private lands. The Farm Bill is not a single piece of legislation, but a dynamic series of acts that include new programs or revises existing ones.

Some of the most effective conservation provisions of the Farm Bill are the Conservation Reserve Program (CRP), the Wetland Reserve Easements (WRE), and the Environmental Quality Incentives Program (EQIP). These provisions directly impact wildlife habitat and help foster healthy natural communities, including upland game bird, waterfowl and big game populations.

Within the Farm Bill effort the CRP has been the single largest contributor (24.2 million acres) in securing upland bird habitat, as well as native grass food sources and cover for white-tailed deer, mule deer, and elk.[47] The WRE has restored over 2.7 million acres of wetlands since its inception in 1990.[48] EQIP is a cost-effective program that has helped install a variety of fish and wildlife improvement projects on approximately 12 million acres under 40,000 different contracts.[49]

Landowners enrolled in these programs receive incentives for land conservation practices. These include native grass establishment, wetland restoration, and forest restoration or enhancement. These programs gained traction in the 1980s, partly in reaction to the prevailing farming practices of the 1970s, when increasingly cultivated fence row to fence row to maximize production—a practice that both removed what native habitat was left for wildlife and had a detrimental effect on soil quality. Each Farm Bill has a name and an average five-year shelf life before needing to be reauthorized by Congress.

The 2014 Farm Bill, which reduced the federal deficit by $23 billion, consolidated twenty-three existing conservation programs into thirteen programs, while strengthening tools to protect and conserve land, water, and wildlife. By streamlining programs, the Farm Bill provides added flexibility and ensures conservation programs aim to work for producers and wildlife in the most effective and efficient way possible.

Other important private lands policies include conservation easements through the federal tax code, environmental asset mitigation programs such as those for wetlands and streams, and the US Fish and Wildlife Service's Partners for Wildlife Program. Additional but not fully developed incentive-based programs are those for carbon, water quality, and threatened and endangered species. The Partners for Fish and Wildlife Program, established in 1987, is the primary mechanism for conservation delivery on private lands. Authorization for the program comes from the Fish and Wildlife Act of 1956 (16 U.S.C. § 742a et seq.), which gave the Fish and Wildlife Service broad statutory authority to enter into voluntary agreements with nonfederal government entities, including private landowners. Through these agreements, the program provides assistance to landowners interested in improving fish and wildlife habitats on their properties.[50] As of February 2013, the program has worked with more than 45,000 landowners and 3,000 conservation partners to restore over 1 million acres of wetland habitat, 3 million acres of upland habitat, and 11,000 miles of streams.[51]

The Partners Program is aimed at restoring and enhancing habitat for federal trust resources, which include important and imperiled habitat types and their associated species of concern: migratory birds, threatened and endangered species, interjurisdictional fish, and other species of conservation concern. The program emphasizes restoring degraded wetlands, native grasslands, streams, riparian areas, longleaf forests, and other habitats to their original condition.[52]

The program's philosophy is to work proactively with private landowners for the benefit of both declining federal trust species and the interests of the landowners involved. Usually, a dollar-for-dollar cost-share is achieved by working with landowners and a host of nationally based and local entities (e.g., federal, state, and local agencies, soil and water conservation districts, and private conservation organizations). Landowners make a commitment to restore the habitat for the life of the agreement (at least ten years) and otherwise retain full control of the land.[53]

## Major Conservation Organizations in Conservation Policy

Many of the early conservation organizations were also active in establishing other conservation organizations thus strengthening conservation programs overall. For example, the Boone and Crockett Club worked to establish such organizations as the New York Zoological Society, the Camp Fire Club of America, the National Audubon Society, the American Wildlife Institute, the Save the Redwoods League, Ducks Unlimited, the North American Wildlife Foundation, the National Wildlife Federation, and more recently, the highly effective American Wildlife Conservation Partners. Other effective organizations were also created. Below are a few of them.

In 1906, the National Association of Audubon Societies for the Protection of Wild Birds and Animals was incorporated, officially uniting state groups that had sprung up and established a strong national voice for bird conservation. In 1940, the National Association of Audubon Societies for the Protection of Wild Birds and Animals changed its name to the National Audubon Society. Also concerned with protection, in 1919, the National Parks Association was founded to promote the welfare of the National Park

System and safeguard high standards in the development of national parks. The Izaak Walton League was founded in 1922 to advocate for preserving wetlands, protecting wilderness, and promoting soil and water conservation.

The American Wild Fowlers was founded in 1927 as an offshoot of the Boone and Crockett Club, specifically for game bird management. It later became Ducks Unlimited. The Wilderness Society was founded in 1935 to protect America's last remaining wilderness and inspire Americans to care for wild places. The Environmental Defense Fund was founded in 1967 by a small group of scientists that set out to ban the pesticide DDT.

## Conclusion

Over the last 150 years many individuals and organizations helped establish the conservation policies of North America. This legacy was built upon legislative and institutional innovations critical to the movement, including the creation of the national forest system, the national park system, and the national wildlife refuge system, as well as the federal agencies to oversee those systems; the establishment of modern-day game laws; and the institution of systems for environmental protection and for the conservation of private lands.

This impressive system of hunting and conservation will not sustain itself without solving the many challenges it faces. Land management, hunting, and fishing are being distorted by urbanization, demographic changes, economic demands, political demands, and our crowded, fast-paced lives. Nonetheless, although some may not be cognizant of their importance, our natural resources are vital to all citizens, and each person has a vested interest in conservation. Individuals and groups, laypersons and professionals, throughout the political spectrum, are all stakeholders. However, even among those who are fully aware of their stake, opinions and motivations about the best course of action for a given issue are as diverse as the citizenry itself and are impacted by a virtually limitless pool of variables. We face a moving target and must continually reevaluate both fact and perception and adjust plans and expectations.

As we advance conservation policy for the next generation, we must find ways for all of us, as diverse agencies, organizations, and individuals who don't always agree, to work together as coalitions to address common challenges and opportunities. We must work across political party lines, get to know our elected officials, their record on conservation issues, and help them understand why supporting these issues is politically important.

As concerned citizens and conservation professionals, we all enjoy being in the field. However, if we are to have a future in which wildlife are an important part of North America, we must have, and utilize, expertise in other areas such as policy and communications, political and people skills, leadership and management tools, adaptive and crisis management strategies, and organizational and team-building proficiency.

Except for the late 1800s, at no other time in history has conservation been at such a crossroads in North America where we have the capacity to still make a significant positive impact on it. The challenges to them are great, but with the historic achievements and the financial, political, and scientific assets that it has, North America is in a unique position to successfully address them.

North America has developed a system of conservation laws and institutions that is remarkable not only for its successes, but for its collaborations. Canada and the United States have the right to celebrate how far conservation has come while being challenged to protect this investment and advance it for future generations.

*NOTES*

1. A. Leopold, *Game Management* (New York: Charles Scribner's Sons, 1933), 7.
2. C. Bambery, "Equity in Access to Hunting and the Right to Bear Arms," *Wildlife Mississippi* 3 (2010): 16–20.

3. W. Threlfall, "Conservation and Wildlife Management in Britain," in *Wildlife Conservation Policy*, ed. V. Geist and I. McTaggart-Cowan (Calgary: Detselig Enterprises, 1995), 27–74.
4. Bambery, "Equity in Access."
5. Bambery, "Equity in Access."
6. Threlfall, "Conservation and Wildlife Management."
7. Bambery, "Equity in Access."
8. Bambery, "Equity in Access."
9. Bambery, "Equity in Access."
10. Bambery, "Equity in Access."
11. B. Ballinger and B. Bordelon, "A Brief History of White-Tailed Deer Management," *Wildlife Mississippi* 3 (2013): 16–19.
12. Ballinger and Bordelon, "A Brief History."
13. Bambery, "Equity in Access."
14. Bambery, "Equity in Access."
15. Bambery, "Equity in Access."
16. Bambery, "Equity in Access."
17. Congressional Sportsmen's Foundation, "Right to Hunt, Fish and Harvest Wildlife," *Congressional Sportsmen's Foundation, Policies*, November 16, 2016, http://congressionalsportsmen.org/policies/state/right-to-hunt-fish.
18. V. Geist, "North American Policies of Wildlife Conservation," in *Wildlife Conservation Policy*, ed. V. Geist and I. McTaggart-Cowan (Calgary: Detselig Enterprises, 1995), 77–129.
19. Ballinger and Bordelon, "A Brief History."
20. V. Geist, *Elk Country* (Minnetonka, MN: NorthWord Press, 1991), 19.
21. W. H. Turcotte and D. L. Watts, *Birds of Mississippi* (Jackson: University Press of Mississippi, 1999), 1–2.
22. J. Organ and S. Mahoney, "The Future of the Public Trust: The Legal Status of the Public Trust Doctrine," *Wildlife Professional* 1 (Summer 2007): 18–22.
23. Organ and Mahoney, "The Future of the Public Trust."
24. Organ and Mahoney, "The Future of the Public Trust."
25. Geist, *Elk Country*, 21.
26. V. Geist, "The Club's Legacy: A Continental System of Wildlife Conservation," *Fair Chase* 15, no. 4 (Winter 2000): 15–17.
27. Geist, *Elk Country*, 21.
28. Dictionary.com, s.v. "conservationist," https://www.dictionary.com/browse/conservationist?s=t.
29. United States Fish and Wildlife Service, "Lacey Act," *United States Fish and Wildlife Service, International Affairs*, July 8, 2012, https://www.fws.gov/international/laws-treaties-agreements/us-conservation-laws/lacey-act.html.
30. S. George Bloomfield, ed., *Impertinences: Selected Writings of Elia Peattie, a Journalist in the Gilded Age* (Lincoln: University of Nebraska Press, 2005), 285.
31. D. Duncan and K. Burns, *The National Parks: America's Best Idea* (New York: Knopf Doubleday, 2011), 2.
32. Duncan and Burns, *The National Parks*.
33. J. F. Reiger, "Grinnell, George Bird," *American National Biography*, http://www.anb.org/view/10.1093/anb/9780198606697.001.0001/anb-9780198606697-e-1300661.
34. J. L. Cummins and B. Bordelon, "Famous Conservationists," *Wildlife Mississippi* (Summer 2013): 22–24.
35. US Department of the Interior, National Park Service, "The Department of Everything Else, The Conservation Movement," *Park History*, May 17, 2001, https://www.nps.gov/parkhistory/online_books/utley-mackintosh/interior6.htm.
36. Chapultepec, Inc., "Theodore Roosevelt," Almanac of Theodore Roosevelt, June 26, 2005, http://www.theodore-roosevelt.com/trbioqf.html.
37. Cummins and Bordelon, "Famous Conservationists."
38. Chapultepec, Inc., "Address to Citizens of Dickinson, July 4, 1886," Almanac of Theodore Roosevelt, July 17, 2010, http://www.theodore-roosevelt.com/images/research/txtspeeches/dickinson4july1886speech.pdf.
39. J. M. Long, M. S. Allen, W. F. Porak, and C. D. Suski, "A Historical Perspective of Black Bass Management in the United States," in *Black Bass Diversity: Multidisciplinary Science for Conservation*, American Fisheries Society Symposium 82, ed. M. D. Tringali, J. M. Long, M. S. Allen, and T. Birdsong (Bethesda, MD: American Fisheries Society, 2015), 99–122.
40. US Fish and Wildlife Service, "North American Waterfowl Management Plan: A Model for International Conservation," 2016, https://www.fws.gov/birds/management/bird-management-plans/north-american-waterfowl-management-plan.php.
41. J. F. Organ, "Ding's Brilliant Idea," *Fair Chase* 30, no. 4 (Winter 2014): 36–38.
42. Organ, "Ding's Brilliant Idea."
43. Organ, "Ding's Brilliant Idea."
44. Organ, "Ding's Brilliant Idea."
45. Organ, "Ding's Brilliant Idea."
46. B. Ballinger, A. Boyles, and J. L. Cummins, "Farm Bill Conservation Programs: A Handbook for Lower Mississippi River Landowners," *Wildlife Mississippi*, December 16, 2016, https://www.wildlifemiss.org/Education/PDF/LMR%20FARM%20BILL%20CONSERVATION%20PROGRAMS_12.21.16_reduced.pdf.
47. US Department of Agriculture, Farm Service Agency, "Conservation Reserve Program," USDA Programs and Services, Conservation Programs, June 1, 2018,

https://www.fsa.usda.gov/programs-and-services/conservation-programs/conservation-reserve-program/index.
48. US Department of Agriculture, Natural Resources Conservation Service, "Wetlands," USDA Natural Resources Conservation Science, August 5, 2011, http://www.nrcs.usda.gov/wps/portal/nrcs/main/national/water/wetlands/.
49. US Department of Agriculture, Natural Resources Conservation Service, "Wildlife Habitat Incentive Program," USDA Natural Resources Conservation Science, March 8, 2011.
50. B. Noonan, "Partners for Fish and Wildlife Program," 8th International Wildlife Ranching Symposium, Colorado State University Libraries, 2007, https://mountainscholar.org/handle/10217/86338.
51. V. Chervanka, V., "Senator Inhofe Honored with Private Lands Stewardship Award," United States Fish and Wildlife Service, Partners for Fish and Wildlife Program, April 13, 2018, https://www.fws.gov/partners/.
52. Noonan, "Partners."
53. Noonan, "Partners."

# The Landscape Conservation Movement

William Porter and
Kathryn Frens

*Early in the North American conservation movement, there arose a vocal contingent advocating for legislated landscape protection. This introduced a strong preservationist element to the North American agenda, which would eventually appear to many as being different from, or even opposed to, the hunting-driven utilitarian view on which the North American Model was founded. This chapter explores the historical circumstance that gave rise to landscape conservation efforts. It describes the resulting legacy in terms of institutions, such as national parks, wildlife refuges, national forests, and national monuments and their considerable relevance to the Model, as well as discussing the wide resonance of the Landscape Conservation Movement across social and political divides.*

## Introduction

Washington, DC, 1872. Almost seven years have passed since the end of the Civil War. Reconstruction is underway in the South, the West is being "won," and the eastern United States is staring down the barrel of the Gilded Age. High society is talking about the newly opened Metropolitan Museum of Art in New York City. Members of the general public have recently enjoyed the first professional baseball game. Criminals such as Jesse James and his gang are robbing banks and stagecoaches around the Midwest. And President Ulysses S. Grant is signing Yellowstone National Park into being. Today, Yellowstone is known for its wildlife, including the enormous herd of free-ranging bison. However, in 1872 in Washington, nobody is interested in the bison. Instead, they want to be sure that the spectacular geological features of the area are preserved for the enjoyment of future generations. By signing his name, President Grant introduces a new American innovation to the world: the national park.

The Gilded Age in the United States was a time of mind-boggling change. Known for its unprecedented concentration of wealth in the hands of a few families, Gilded Age society was shifting away from the traditionally American economy of the independent farmer to a manufacturing economy, with consequences for every area of life. Goods that were previously made by specialist tradespersons were now turned out by factories in record numbers. News arrived by telegraph from all over the world, and cities were lit at night with electric lights. The end of the Civil War had recently left large numbers of ex-soldiers both traumatized and out of work. Newly freed African Americans claimed land for farms or moved to the industrial Northeast to find jobs. They were joined there by hundreds of thousands of new immigrants, who labored for little pay and in dangerous conditions in meatpacking plants, cotton mills, and shipyards. Europe was in the throes of an intellectual upheaval: Charles Darwin published his masterwork *On the Origin of Species* in 1859, just eleven years after Karl Marx's *Communist Manifesto* first hit the presses. Both Darwin's scientific approach to nature and the increasing popularity of collectivist thought would help to shape the conservation movement in the United States.

Changes to the national transportation system during the Gilded Age brought American entrepreneurs to a land of seemingly unlimited resources. While the Erie Canal had been opened in 1825, the completion of the transcontinental railroad in 1869 allowed access to the natural resources of the Great Lakes, the Ohio River Valley, and the West. In the aftermath of the Civil War, economies in the eastern United States were booming, demanding enormous amounts of food, fiber, and minerals. The more than 100 million acres of forests stretching from Maine to Minnesota looked like an endless supply of wood to eastern businessmen; the flocks of passenger pigeons looked like infinite dinners.

The improved transportation system enabled the evolution of the market hunter from a local supplier to a new type of small businessman who spent his time on the Great Plains, killing bison or passenger pigeons to ship back east on the trains. Speedy (and refrigerated) transportation opened a market for game meat, buffalo robes, furs, and feathers, with drastic consequences for wildlife. Until now, it had not occurred to Americans that an entire species could go extinct, but in the 1870s, passenger pigeons and bison looked to be on the way there. Westerners and Easterners alike began to consider the idea of resource scarcity, a necessary first step toward the idea of wildlife conservation.

The pace of exploitation of game and other resources amazed the individuals who saw it. Railroads carried scientists, land surveyors, and the first wealthy tourists into the wild lands of the West. Early ecologists, most prominently George Perkins Marsh, began to write about the connectivity of land, nature, and human society. They warned of the unintended consequences of such large-scale exploitation of natural resources, which included the loss of what modern scientists call "ecosystem services."

By the 1880s, the water level of the Erie Canal had fallen noticeably, hampering transportation of goods and people along this essential travel corridor. Businessmen complained, and scientists pointed to the Adirondack Mountains of northern New York, which had been extensively logged in the previous decades. Previously, the forests of the Adirondacks had acted as a giant sponge, soaking up heavy rainfall and releasing it slowly. The loss of trees reduced the capacity of the sponge and disrupted the hydrology that fed the Erie Canal. Officials from local and state governments realized that protecting forests in the Adirondacks was in their economic best interests, and they acted quickly. In 1885, the New York State Legislature established the Adirondack Forest preserve, protecting large areas from future exploitation. The intention was to ensure the Erie Canal's water supply by protecting the land.

In 1890, the last shipment of bison hides has gone east, and William Hornaday estimates the free-ranging bison population at eighty-five individuals. He had traveled to Montana to collect specimens so that future visitors to the US National Museum (his employer) would know what bison looked like, after their imminent extinction. That same year, a census of all people in the United States signals the end of an era: President Benjamin Harrison declares the American frontier closed. The country of Daniel Boone, Sacajawea, and Crazy Horse is now considered settled. The Wild West begins to calm down. The ongoing war against Native Americans has ended with the massacre at Wounded Knee. The shootout at the OK Corral is now history, and Wyatt Earp is running a restaurant in San Diego.

The first century of American history had been defined by the frontier. Upon arrival in a land much less densely populated than they were used to, the first Euro-American settlers immediately began clearing land. In the Puritan faith, to which many of these early immigrants adhered, wilderness was a source and a symbol of evil and chaos. By clearing and planting crops on wild forest land, the Puritans believed themselves to be doing work ordained by God, bringing order and productivity out of unredeemed chaos. Later waves of immigrants were just as skeptical of wilderness, often for somewhat more practical reasons. They cut their farms out of forest, or into the mat of tough prairie sod, or on thin-soiled hillsides. Every year, the native vegetation encroached, and the farmers fought it back. To most of them, the ideal landscape was pastoral: orderly rows of corn, an orchard, perhaps sheep dotted across a green meadow. Wild land was the adversary that threatened their lives and livelihoods.

The closing of the frontier probably seemed purely symbolic to many in the West. While land was much harder to come by, many families in remote areas continued to scratch out a living from wild land. Farmers fought constantly to extract new agricultural land from its wild state. But in wealthy Eastern circles, the perception of wilderness had begun to change. Manufacturing and transportation had brought about a larger upper class, which had time and money to approach the natural world as more than the sum of its resources. For this new leisure class, mountains were scenery rather than obstacles. Forests produced air cleaner than what they breathed in crowded cities polluted with the residue of burning coal. Trees could be admired without any thought of cutting them down for lumber. For perhaps the first time, some in American society were struck by two realizations: wilderness was both beautiful and useful, and it was disappearing. These realizations were the fundamental ingredients to what would soon become the drive to preserve the wilderness that remained.

Some people, mostly wealthy Easterners, began to seek out the wilderness experience. They bought landscape paintings of mountains and forests seemingly untouched by human hands. They escaped to the Adirondack Mountains in the worst heat of summer. They wrote and talked of wilderness in a new way, reviving the Transcendentalist thinking first popularized by Thoreau and Emerson. Thoreau wrote that wilderness was important to the "intellectual nourishment of civilization," a decidedly minority view when he published *Walden Pond* in 1854. In the late 1800s, however, the new upper class embraced this idea and began to advocate for wilderness preservation on the grounds that wilderness was necessary for the spiritual wellbeing of humans, especially city-dwellers.

However much they might have admired the mythology of pioneers and mountain men, members of the new leisure class were not interested in experiencing wilderness as their forebears had. They built "great camps" in the Adirondacks, rustic-looking mansions complete with electricity and plumbing, and brought their servants on summer getaways. Even Theodore Roosevelt, a more accomplished outdoorsman than most of his class, employed guides and porters on his hunting expeditions. Wilderness was no longer an objective physical reality, but an experience defined by the participants.

The origins of conservation, and much of what we see in the North American Model, are traced to these early ideas about setting aside land. Wildlife was important and would become central to the North American contribution to conservation. But it was wild landscapes that captured the hearts and minds of many in American society and spurred the revolutionary idea of conservation. The cradle of this revolution is often considered Yellowstone, but two other landscapes also share this distinction: the Adirondack Mountains of northern New York State and Yosemite in California. Their stories are essential to understanding early conservation.

## The Adirondacks: Forever Wild

The Adirondack Mountains of northern New York, a 10,000-square-mile region of mountains, lakes, and forests, were the first exposure most Americans had with gross exploitation of natural resources and the value of wilderness. Due to the inhospitable climate and terrain, neither indigenous Americans nor early Euro-Americans built settlements there. It was only when the demand for lumber made logging commercially viable that the Adirondack region was permanently changed by human activity. The Erie Canal and the new railroads enabled more efficient transportation of timber to faraway places, and new technology such as the two-man crosscut saw made cutting down trees easier and faster than ever before. Already by the 1850s, naturalists and land surveyors warned of the damage done by unsustainable logging in the area. Timber companies bought land, removed most or all of the trees, then let the land revert to state ownership by refusing to pay taxes.

The Adirondacks' proximity to New York City, Philadelphia, and Boston helped attract the attention of soon-to-be conservationists and lovers of wilderness, as well as businesspersons interested in shipping on the Erie Canal. In 1885, New York designated all public property in the Adirondacks as the Adirondack Forest Preserve, and seven years later, added unprecedented protections for the preserve to the state constitution in what would be the first large area of land declared Forever Wild:

> The lands of the state, now owned or hereafter acquired, constituting the forest preserve as now fixed by law, shall be forever kept as wild forest lands. They shall not be leased, sold or exchanged, or be taken by any corporation, public or private, nor shall the timber thereon be sold, removed or destroyed.[1]

Railroads that helped exploit the forest also brought increasing numbers of travelers and vacationers. The Adirondacks were becoming a premiere tourist destination for many of the East Coast's elite. Industry tycoons bought large expanses of land and built expansive summer homes. While game animals had been historically scarce in the area, the newly regenerating forests promoted rapid growth of populations of deer, grouse, and other game animals. These large homes became known as great camps, they resembled English country estates and providing all the comforts for the wealthy elite but removed from the increasingly crowded and polluted cities. The Adirondack wilderness was considered a healthier environment for children and sick people. In 1885, the country's first tuberculosis sanatorium, or treatment center, opened in Saranac Lake, a village in the heart of the Adirondack Mountains.

Released from the constraints of making a living off the land, the wealthy took up hunting as a sport, and in doing so drastically changed how game animals were managed. Sport hunters were concerned with hunting ethics and the principles of "fair chase," a set of standards that opposed the methods of subsistence and market hunters, who were more concerned with the success of their hunting than with how they hunted. Over time, as market hunting was outlawed and the number of subsistence hunters declined, these standards became ingrained in American culture, and thus became integrated into the North American Model of Wildlife Conservation.

New Yorkers understood that unfettered resource exploitation had come very close to causing severe and perhaps irreparable damage to the mountains and forests that many of them had grown to love. This near-disaster sparked an interest in more sustainable approaches to forest management and heralded the introduction of science as a cornerstone of the notion of conservation. Such a use of science to govern wise stewardship of natural resources was novel to a society that had very recently thought of natural resources as limitless. Theodore Roosevelt, who had seen the ravages of exploitation in the Adirondacks, had trained in science at Harvard. When he met Gifford Pinchot, a forester who was steeped in the science of silviculture, their joint efforts initiated the broader application of science to management of natural resources. Thus, science became centrally important to the North American Model of Wildlife Conservation.

Inclusive conservation would later be recognized as one of the most important legacies of the Adirondack story. While the designation of Forever Wild applied to the trees on land in public ownership, hunting and fishing were not prohibited. Not quite a century later, New York State established the Adirondack Park, a 6-million acre region composed of a checkerboard of public and private lands. About half of the land was publicly owned. The remainder was owned by myriad timber companies, other businesses, and private individuals. Across all of this land, both public and private, the people of New York invoked a suite of land-use regulations of unprecedented stringency. The intent was to promote not only a healthy wilderness ecosystem but also a vibrant human economy. People were explicitly included.

## Yellowstone: Natural Wonders

The first known Euro-American to have visited the area that is now Yellowstone National Park described its hot springs and geysers in such outlandish terms that his listeners back East thought he was making it up. It was only when later visitors confirmed his descriptions of spectacular geological activity that larger society became interested. But Yellowstone's history goes back much farther. Obsidian was quarried from cliffs in the park beginning more than 11,000 years ago and dispersed via trade to places as far away as Ohio. People from the Shoshone, Bannock, Nez Perce, Flathead, Crow, and Cheyenne tribes came to the area regularly to hunt, gather, and use the hot springs for curing hides. When the area was designated a national park, these tribes were considered intruders, and their traditional uses of the land were prohibited. In contrast to the Adirondacks, Yellowstone was designed to exclude people from living in the park.

Yellowstone's natural wonders were what first drew the interest of American society. How best could society protect the fantastic geological features from the crass commercial development that was sure to come with the railroads? Moreover, the vast expanse of the park's land would soon become an equally valuable asset, supporting wildlife in numbers almost impossible in the rest of the country. Indeed, wildlife came to the fore quickly. When several park superintendents ignored poaching in the park, the Army was called in to help enforce the anti-poaching laws. These laws were aimed particularly at protecting bison, which still existed only in captivity and in Yellowstone. Locals who depended on hunting for their food cheerfully ignored the anti-poaching laws, and juries of their peers almost never convicted them; they had always hunted elk, bison, and deer whenever and wherever they could, and they believed that no government should be able to stop them.

The problem, as the locals saw it, was the market hunters. Game in other places was dwindling, but market hunting in Yellowstone was risky, especially after the park hired local "scouts" to find and arrest poachers. Career hunters eluded capture in a number of increasingly creative ways, the worst of which (to locals) was taking only the animal's most valuable parts—the head of a bison, the teeth of an elk—and leaving the rest. The neighbors of such poachers often knew exactly what they were up to and passed tips to park authorities, telling them where to find poachers' hideouts and caches of game. William Binkley of Jackson Hole, a notorious market hunter, was the target of his neighbors' anti-poaching efforts for several years. In 1899, Jackson Hole residents took up a collection to hire another game warden for the area with the intent of controlling illegal activity by Binkley and other commercial hunters. Three years later, they formed a "Game Protective Association" with the same goal. Finally, in 1906, a small group of residents met and debated whether to lynch Binkley outright. Binkley criticized his neighbors' hypocrisy, saying that many of them had also hunted illegally, but he failed to distinguish between subsistence and market hunting. Residents of Jackson Hole considered hunting for food to be an inalienable right.

In a sense, Yellowstone, like the Adirondacks, was an ongoing experiment in search of a clear philosophical foundation and day-to-day management principles. If there were philosophical underpinnings, they were rooted in a society now romanticizing the rough-and-ready ethos of the Old West. There were no principles drawn from science or experience to follow. Essentially, they were making up whatever they thought needed to be done as they went along. Much of the management that developed arose from adapting to each new challenge. Yet the heart of this emerging philosophy of conservation and the daily experiment in implementing management was captured in an extraordinarily prescient phrase that is carved in the stone of the Roosevelt Arch at the north entrance to the park: "For the Benefit and Enjoyment of the People."

## Yosemite: The Cathedral

Onto this over-exploited, cowboy-ridden romanticized landscape stepped a very different kind of wilderness advocate, one whose message drew devotees from across classes and regions and fostered a passion for wild places. John Muir had the eloquence of an Old Testament prophet and the beard to match. As a young man, he hiked from Indianapolis to Florida and did a stint draft-dodging in southern Ontario. Yet it was California that changed his life and propelled him onto the national stage. For several years, he lived in a cabin in the Yosemite Valley, which he referred to as "God's big show." Yosemite's towering rock formations and waterfalls had been astounding visitors for decades, but Muir's writings ascribed to the scenery a spirituality that previous visitors had not. He described the valley as "full of God's thoughts, a place of peace and safety amid the most exalted grandeur and enthusiastic action, a new song, a place of beginnings abounding in first lessons of life . . . with sermons in stone, storms, trees, flowers, and animals brimful with humanity."[2]

The concept of conservation took on richer dimensions in Yosemite. Whereas the conservation of the Adirondacks was based on pragmatic ideas, including the goals of better hydrology and better health, and whereas the concept of protecting natural wonders motivated setting aside Yellowstone, it was Yosemite that tapped straight into the spiritual basis of conservation. The religious nature of Muir's writings about Yosemite intrigued many who read them. Americans were used to thinking of a God who brought order out of the wilderness, not a God who lived in the wilderness itself. And yet, the idea resonated with the increasingly urban population. Yosemite was a place of wonder, whether experienced through a visit or through the eyes of the many artists who painted or photographed its landscapes. The spiritual connection grew to be paramount for many people, as the devout ascribed the wonder of this landscape to its creator.

"In God's wilderness lies the hope of the world," Muir wrote, "the great fresh unblighted, unredeemed wilderness."[3] Muir challenged the pioneer vision of wilderness as an evil, disordered force and asserted that the "unredeemed" nature of wilderness is what makes it good. In just a few short human generations, wilderness in the American mind had gone from "red in tooth and claw" to "the preservation of the world."

Thanks to Muir's writing, Yosemite National Park was created in 1890 and was the first piece of land explicitly protected for the value of its wilderness alone. In reality, the Yosemite Valley had been seasonally inhabited for centuries, but the native residents had been removed by the 1850s. The damage caused by sheep grazing had been one impetus for the protection of Yosemite, and Muir worried that the interests of farmers and what he called "hoofed locusts" would eventually win out over land protection. He and several other lovers of wilderness formed the Sierra Club in 1892 in order to "do something for wildness and make the mountains glad."[4] While their advocacy for the spiritual and aesthetic value of wilderness had broadened the appeal of wilderness conservation, it also set the stage for a schism among advocates of conservation that would be played out over the next century.

## The Roosevelt Presidency

September 14, 1901. The Spanish-American War has been over for less than three years, and the US government is busy figuring out how to run its new territories: Hawaii, the Philippines, Puerto Rico, and Guam. Mount Ranier in Washington is the newest national park. There are twenty-three bison in Yellowstone, with a few hundred more in private herds, and the government issues a new $10 note with a bison on the front. President William McKinley is shot by an anarchist at the Pan-American Exposition on September 6, but seems to be recovering under the care of his doctors. So Vice President Theodore Roosevelt joins his family on vacation in Newcomb, New York, in the Adirondack Mountains. Several days later, a messenger reaches him on Mount Marcy with news that the president is dying. Roosevelt's hair-raising ride from Newcomb to North Creek—down a mountain in a horse-drawn buggy, at night, on rough roads, in the rain—becomes the stuff of Adirondack legend.

Theodore Roosevelt's presidency is often regarded as the heyday of the Progressive Era, but it was even more notable for the contributions that Roosevelt made to the land conservation movement. Roosevelt had long been interested in what he called "the strenuous life," taking trips out West to hunt big game and to camp in Yosemite with John Muir. Years before he became president, he co-founded the Boone and Crockett Club, the country's first wildlife conservation organization, which was dedicated to the conservation of game species and their habitats. Like Boone and Crockett, conservationists were starting to recognize the commonalities between wilderness and wildlife conservation, and the two were beginning to be integrated.

In his seven years in office, Roosevelt set aside more public land than anyone at the time could possibly conceive. With Gifford Pinchot, he designated the first national forests and founded the US Forest Service to manage them. This management, which included harvesting timber at a rate that would not permanently harm the forest, was seemingly a compromise between wilderness and development advocates. Not everyone was happy with this bargain, though. Muir and his wilderness advocates had hoped that the national forests would remain wild, and they broke with Pinchot completely when Pinchot decided to allow Muir's old enemy, sheep, to graze in the national forests. Pinchot's doctrine of "wise use" was ascendant, and timber and game harvesting would become the rule, rather than the exception, on most public lands. Muir would spend the rest of his life fighting for the protection of wilderness—and mostly losing. Near the end of his life, he mobilized wilderness lovers across the country in the public relations fight against building the Hetch Hetchy Dam in Yosemite National Park. Roosevelt, torn between Muir's "preservationists" and Pinchot's "conservationists" (who supported the dam), waffled. In the end, he authorized construction of the dam, believing that the public would not accept the preservation of wilderness at the expense of San Francisco's need for water.

On the surface, the American system of parks and reserves echoed the Old World system, in which game animals belonged to large estates that were kept for hunting. The important difference reflected the best of American ideals. Instead of belonging to a wealthy landholder, the American national parks and other reserved lands belonged to "the people." In the highly capitalistic United States, common ownership of property was highly unusual. To this day, the United States is one of the few developed countries without a national bank or nationalized utilities, but still has one of the most extensive public land systems in the world. This juxtaposition hints at the importance of wilderness and wildlife to American culture.

Wildlife management would reflect lessons learned from the wilderness movement. In another principle captured in the North American Model of Wildlife Conservation, wildlife would be collectively managed by the government as a public trust, and everyone would have the right to hunt, subject to individual state regulations. The sale of wild game products (e.g., commercial or market hunting) was prohibited. If this system did redistribute the benefits of wildlife mostly to that segment of the public that could afford recreational hunting while stringently regulating subsistence hunting, it also allowed more centralized and integrated management of land and wildlife.

A Harvard-educated biologist, Roosevelt also embraced scientific management as part of his conservation agenda. Wildlife had long been the domain of naturalists, who were concerned primarily with the description and classification of plants and animals. However, two influential Britons would revolutionize the science of the natural world. Charles Lyell, a geologist, and Charles Darwin, a biologist, each developed theories integrating the naturalists' descriptive work with ongoing processes in their respective fields. By recognizing that both living and nonliving systems evolve under the influence of natural processes, these scientists provided the foundation for the study of ecology and, ultimately, for wildlife and land management. Wildlife management as a science did not exist yet, and wouldn't exist until the 1930s, so scientific land management in the United States was pioneered by foresters. The goal of these early forest scientists was to find ways to harvest timber without destroying forests—a concept that we would recognize today as "sustainability" and one that Roosevelt advocated in his day. Taking a long view, he wrote:

> Defenders of the short-sighted men who in their greed and selfishness will, if permitted, rob our country of half its charm by their reckless extermination of all useful and beautiful wild things sometimes seek to champion them by saying that "the game belongs to the people." So it does; and not merely to the people now alive, but to unborn people. The "greatest good for the greatest number" applies to the number within the womb of time, compared to which those now alive form but an insignificant fraction. Our duty to the whole, including unborn generations, bids us to restrain the unprincipled present-day minority from wasting the heritage of those unborn generations. The movement of conservation of wild life and the larger movement for conservation of all our natural resources are essentially democratic in spirit, purpose, and method."[5]

While Roosevelt and others understood the power of abstract ideas, they were ultimately pragmatists. They recognized that conserving wildlife for future generations was about the connection of wildlife to the land. Lands set aside within the national forests provided habitat for wildlife; federal enforcement of protection from illegal hunting meant more game for future generations. Indeed, nearly three generations before "biodiversity" came into American discourse, Roosevelt recognized its value and in 1902 established the first national wildlife refuge in Pelican Island, Florida, to protect an extraordinary diversity of birds. By the end of his administration, Roosevelt had created more than fifty national wildlife refuges in seventeen states and US territories, most by means of executive action, sometimes over the objection of Congress.

November 1909. Roosevelt has left the presidency and the country for an African safari to collect specimens for the American Museum of Natural History. Henry Ford's Model T is three years old, and the Chicago Cubs are in the first year of their World Series drought, having defeated Ty Cobb's Detroit Tigers the previous year. Thirty-four captive bison are released into a new bison sanctuary. In the Apache National Forest of Arizona, a young forester kills an old wolf in a river valley. The flash of an idea in the wolf's dying eyes will push wilderness conservation into a new era.

Following their interpretation of the wise use policy, many government agencies attempted to extirpate wolves, cougars, and other predators in order to protect game populations and domestic livestock. The value of wilderness land might have been accepted, but only as far as it served some human need. The science of ecology was still young, and managers had no idea what a herd of deer with no predators could do to a forest. Aldo Leopold, the wolf-hunting forester, described it years later:

> I was young then, and full of trigger-itch; I thought that because fewer wolves meant more deer, that no wolves would mean hunters' paradise. But after seeing the green fire die, I sensed that neither the wolf nor the mountain agreed with such a view. Since then I have lived to see state after state extirpate its wolves. I have watched the face of many a newly wolfless mountain, and seen the south-facing slopes wrinkle with a maze of new deer trails. I have seen every edible bush and seedling browsed, first to anaemic desuetude, and then to death. I have seen every edible tree defoliated to the height of a saddlehorn. . . . I now suspect that just as a deer herd lives in mortal fear of its wolves, so does a mountain live in mortal fear of its deer."[6]

Leopold grew into a wilderness philosopher in the mold of Muir and Thoreau, with the added benefit of an ecological education. He is now considered the father of wildlife management, and his writings on land and wilderness are still considered some of the most eloquent articulations of the importance of wild land to humanity. In fact, the philosophical foundation of the modern wilderness movement has changed very little since Leopold's death in 1948.

The role of science in land and wildlife management continued to expand, providing the insight that all elements of an ecosystem—land, water, wildlife, microbes—are interconnected and each affects the others and the whole system. Still, as Leopold came to recognize, it was not science but philosophy that enabled a paradigm shift in our collective understanding of wilderness. Leopold wrote:

> An ethic may be regarded as a mode of guidance for meeting ecological situations so new or intricate, or involving such deferred reactions, that the path of social expediency is not discernible to the average individual. Animal instincts are modes of guidance for the individual in meeting such situations. Ethics are possibly a kind of community instinct in-the-making. . . . All ethics so far evolved rest upon a single premise that the individual is a member of a community of interdependent parts. . . . The land ethic simply enlarges the boundaries of the community to include soils, waters, plants, and animals, or collectively: the land.[7]

## Conclusion

Throughout Western history, humans had seen land as a commodity to be used—perhaps carefully, sustainably, at times with sound science, but more or less as a means to increase human wellbeing. Leopold's proposal of a land ethic instead welcomed land, including all its component parts, into a broader community that included humans. Thus, the farmer growing windbreaks and grassed waterways to preserve the soil creates habitat for catbirds and quail, woodchucks and deer. This paradigm sees humans not as the masters of land, but as co-equal beings in a relationship with land. Thus, wilderness is not important only for the value of its ecosystem services or its spiritual contribution to humanity; it is intrinsically important in itself, just as individual human beings have intrinsic importance along with the importance of what we contribute to society. Humans are included in the ecosystem; land is included in the community. As Leopold wrote,

> A land ethic of course cannot prevent the alteration, management, and use of these "resources," but it does affirm their right to continued existence, and, at least in spots, their continued existence in a natural state. In short, a land ethic changes the role of *Homo sapiens* from conqueror of the land community to plain member and citizen of it. It implies respect for his fellow-members, and also respect for the community as such.[8]

Leopold's land ethic represents the natural extension of the evolution of conservation as an American ideal. At the center of conservation is an ethic. It is because of this ethic that we as a society behave differently from previous generations through the restraint we apply in our treatment of the land, forests, wildlife—all of nature. We consider prescient the insights of Theodore Roosevelt, Gifford Pinchot, John Muir, and Aldo Leopold. We recognize the epiphany that was captured in the North American Model of Wildlife Conservation. Their insights speak to us about the evolution of public ownership, scientific management, and sustainability. And perhaps, most powerful, the core principle of conservation was at the beginning, and is still today, about the land. The movement that has propelled American conservation to become one of the greatest ideas in all humanity, has its origins in setting aside lands as Forever Wild, in the preservation of the wonders of nature, and in the recognition of the spiritual necessity of nature to the well-being of humans as part of a community of life.

*NOTES*

1. New York Department of State, Division of Administrative Rules, "The Constitution of the State of New York," *New York State Constitution*, 2014, https://www.dos.ny.gov/info/constitution.htm.
2. John Muir, "The Yosemite National Park," *Atlantic Monthly*, August 1899, http://www.theatlantic.com/past/docs/issues/1899aug/muir.htm.
3. Linnie Marsh Wolfe, *John of the Mountains: The Unpublished Journals of John Muir* (Madison: University of Wisconsin Press, 1938), 317.
4. Sierra Club, "Who Was John Muir?" *Sierra Club*, April 9, 2010, http://vault.sierraclub.org/john_muir_exhibit/intro.aspx.
5. Douglas Brinkley, "A Book-Lover's Holidays in the Open," in *The Wilderness Warrior: Theodore Roosevelt and the Crusade for America* (New York: Harper Collins, 2009), vii.
6. Aldo Leopold, *A Sand County Almanac* (New York: Oxford University Press, 1949), 129–132.
7. Leopold, *A Sand County Almanac*, 203–204.
8. Leopold, *A Sand County Almanac*, 204.

James R. Heffelfinger and Shane P. Mahoney

# Hunting and Vested Interests as the Spine of the North American Model

*One of the great achievements and surprising aspects of the early North American conservation movement was the recognition that self-interest must play a role in any long-term, effective strategy. The North American Model, as it emerged, clearly represented a strong, practical approach that would return something of direct value to the citizenry of the day. It was in this frame of reference that hunting, defined as a fair-chase recreational pursuit (in contradiction with market killing), became the cornerstone of the North American approach to wildlife protection and enhancement. Thus, the North American Model presaged, by nearly a century, the sustainable use approach of modern global conservation parlance and practice. This chapter explores the evolution of the North American hunter and describes how the hunting of wild animals for personal use and recreation became, and appreciably remains, a political, economic, and social agency of great importance to North American conservation.*

## Introduction

The North American Model of Wildlife Conservation did not originate as an a priori strategic plan that was carefully laid out and then implemented. Instead, the term is an a posteriori description of the successful conservation paradigm that developed through time as a result of a large collection of laws and institutions and changing public perceptions of wildlife and the broader issues of conservation.

It is perhaps surprising that such a system could develop without a detailed guiding vision from the start. How could the many laws and policies coalesce into a system of conservation that has resulted in the retention of many native wildlife species, including a suite of large carnivores? Certainly, part of the answer lies in the fact that from its inception the North American approach recognized that self-interest must play a role in conservation practice.[1]

In the early years of American settlement, unregulated market killing of wildlife returned a short-term benefit to individuals, but its unsustainability and destruction of wildlife represented detriments to society at large. It was in this context that hunting, defined as the fair chase pursuit of wildlife available to everyone, became the cornerstone of the North American approach to wildlife protection and management.[2] While rising urban elites played a significant role in launching the movement for the conservation of wildlife and habitat, the North American Model reflected a strong practical conservation approach that provided something of direct value—the conservation and restoration of wildlife resources—to the citizenry of the day. Thus, the North American Model presaged by nearly a century the sustainable use approach of many modern global conservation policies, which seek incentives as crucial drivers for successful program implementation.

## From Subsistence to Sustainability

At the early stages of Western civilization in North America, hunting and trapping were important survival skills, and the harvest of wildlife remained one of the leading contributors to the economies of young, growing nations. However, unlike the indigenous cultures they encountered, European settlers armed with rifles easily overharvested certain species of wildlife that had the misfortune of being edible, wrapped in fur and leather, adorned by ornate feathers, or otherwise desirable. Through a long period of unmitigated wildlife slaughter for market and subsistence use came the gradual realization that the rate of killing was unsustainable and something had to change, an awakening strongly influenced by the fates of the passenger pigeon and the American bison.[3] Concerned citizens, led by some of the most prominent hunters and conservationists of the day, developed and promoted a collective sense of stewardship and a conservation ethic.[4] The desire to maintain wildlife populations evolved into a successful system we now recognize as the North American Model of Wildlife Conservation, in which personal interest in sustainable, consumptive use of wildlife is the cornerstone.

This conservation paradigm, built upon the foundation of scientifically regulated harvest, has benefited many species of native wildlife, both hunted and nonhunted. As wildlife practitioners conserved, restored, and managed habitat for game species, they also conserved landscapes that provided habitat for nonhunted species. Indeed, the North American Model helped develop a worldwide recognition of regulated hunting as a demonstrably valuable tool for the conservation and sustainable use of wildlife. At international symposia held in the 2000s in London and Namibia, for instance, participants reaffirmed the importance of hunting to wildlife conservation.[5]

Support for wildlife by hunters and trappers is well documented and arises partly out of self-interest because of the importance of wild harvest to them.[6] Participants accrue many benefits from hunting and trapping, ranging from psychological to physical, from sociological to nutritional. However, the real collective benefit in North America is the stewardship of wildlife and the habitats on which they rely.[7] Hunters have contributed billions of dollars to support conservation of wildlife species, in general, not just those that are hunted.[8] The hunting community has proven itself to be a central pillar of the North American conservation paradigm and is thus responsible for supporting a wide variety of conservation activities that are highly valued by the broad public.

## Conservation through Use

Both the need for and success of the North American Model of Wildlife Conservation have revolved around sustainability of harvest.[9] In its infancy, wildlife management in North America began as a system of harvest limits to stop the rapid decline of wildlife caused by overexploitation to meet commercial meat demands.[10] When unregulated killing was controlled in concert with growing restoration efforts, populations rebounded vigorously.[11] As laws and policies gained effectiveness, programs to restore and manage hunted species gradually emerged. Thus, regulated hunting influenced a bold, experimental system of comprehensive wildlife conservation.

## Determining Harvest Levels and Management Goals

In North America today, state and provincial wildlife agencies use research, population monitoring, adaptive management, emerging technologies, and experience to develop management guidelines and protocols to determine the appropriate level of harvest for game populations. The allowable harvest can be that which is sustainable with no ill effects to the population, or it might be a prescription to reach specific management goals of animal abundance and demography. Social pressures play a role in many cases, as when issues such as animal-vehicle collisions, agricultural crop damage, or hunter preferences influence management goals.

In most wildlife management scenarios, there exists a wide range of management goals that are appropriate, both ecologically and biologically. These goals are normally set through an open and transparent process that allows the public ample opportunity to provide input. Additionally, agencies use scientifically designed surveys to gauge public opinion and preferences.[12] Because wildlife is held in public trust and managed on behalf of all citizens in North America, management consistently reflects the broader public voice. The needs and wants of all citizens are considered in how wildlife is managed. Once goals are determined, proper management involves monitoring wildlife population abundance and demography, as well as harvest-related parameters. Both survey (e.g., abundance, distribution, sex and age ratios) and harvest (e.g., number harvested, age structure, harvest per unit effort) information represent vital data that help biologists manage wildlife populations.[13]

## Managing Animal Abundance

Hunters are critical to the current system of North American wildlife conservation and are the most effective logistical agents for wildlife population management. The early days of North American wildlife management were spent stopping declines of species that humans deemed worth saving, as well as encouraging population growth through limited hunting seasons, sex-selective hunting for some species, daily bag limits, and other restrictions. As law enforcement, habitat restoration and protection, and wildlife management programs grew, however, so did most wildlife populations. The initial protections intended to conserve species actually resulted in an overabundance of some. Biologists, hunters, and the public soon realized that some populations had begun to exceed the carrying capacity of their habitat.[14]

It is often necessary to maintain wildlife populations below environmental and social carrying capacity to reduce die-offs, to provide for more productive populations, to protect habitat, to reduce the spread of disease, and/or to reduce conflicts with humans.[15] In cases where population reduction is the management goal, managers must implement harvest beyond the level at which the population can replace itself in the short term.[16] Population reductions or maintenance at desirable levels are examples of hunters working as partners in wildlife management. Conflicts with humans in the form of vehicle collisions, nuisance wildlife, livestock depredation, as well as the needs of agricultural production and human and domestic animal health and safety, may result in a goal to manage at a social carrying capacity lower than the biological limit of the habitat.[17] For example, the number of deer-vehicle collisions is estimated to exceed 1.5 million every year on US roadways.[18] The related, and tremendous, loss of life and property illustrates the need to effectively managing wildlife abundance.

In recent decades, hunting alone has proven ineffective in controlling certain overabundant populations, most obviously white-tailed deer (*Odocoileus virginianus*).[19] White-tailed deer are a highly adaptable and prolific species that benefit from habitat disturbance, including agriculture, and urban refugia. The challenge of managing their abundance has been exacerbated by declines in hunter retention and recruitment, fewer accessible hunting opportunities, and changing human attitudes, including a reluctance to harvest females and the increasing prevalence of protectionism. This is not an indictment of the effectiveness of hunting as a wildlife management tool in general; rather, it is an example of the social and biological complexities of wildlife management.

The importance of hunting to conservation, in the broad sense, is not tied simply to population control, but is far more complicated. For example, if prey species have to be hunted because predator populations were reduced, why are predators still being hunted?

One must understand that a deer season or duck season it is merely a component—a critical one—of a much larger, successful wildlife conservation model.

## Funding and Implementing the North American Model

Many people enjoy wildlife; however, not everyone is aware of the financial contributions made by hunters, trappers, anglers, and recreational shooters to support sustainable conservation programs. Nor does everyone realize the fundamental role these individuals play in preserving wildlife and wild places.[20] Although many consumptive users know that financial contributions from hunting license sales, as well as a percentage of some outdoor equipment taxes, help pay for wildlife management, they do not always recognize their own key role in conservation on a broader scale.

The economic contributions of consumptive users to conservation activities are immense and are currently unmatched by contributions from any other group. In 2016, $695 million was apportioned to state wildlife agencies in the United States from the excise tax collected on hunting and shooting purchases (Federal Aid in Wildlife Restoration Funds, also known as the Pittman-Robertson Act).[21] Annually, sales of hunting and trapping licenses (more than $600 million) in the United States and private donations by hunters for conservation efforts (more than $300 million) combine with Federal Aid in Wildlife Restoration Funds to raise more than $1.6 billion per year for conservation.[22] Some of these contributions by hunters and other vested interests are voluntary, but most are a requirement of participation. Canadian provinces have also demonstrated significant annual economic contributions from hunting and hunting-related activities.[23] All revenue from hunting and fishing in Canada is added to federal general revenue and then distributed to provincial fish and wildlife agencies.

In addition to institutionalized programs, nongovernmental organizations in both the United States and Canada, often founded and funded by hunters, raise and contribute additional money for habitat acquisition and enhancement, population restoration and monitoring, and specific wildlife research projects. For example, the Wild Sheep Foundation (WSF) has raised and contributed more than $70 million in the last thirty years to activities that benefit wild sheep and other wildlife.[24] Similarly, the Rocky Mountain Elk Foundation has funded 7,400 individual projects protecting or enhancing more than 5.9 million acres of wildlife habitat.[25] Since 1973, the National Wild Turkey Federation (NWTF), in cooperation with state and federal partners, has spent more than $331 million to restore wild turkeys and to conserve more than 16 million acres of habitat.[26]

## Habitat Acquisition, Protection, Restoration, and Enhancement

Land management agencies manage wildlife habitat on millions of hectares of federal and Crown land. Many US states have also purchased wildlife habitat with the proceeds from hunting licenses and taxes on some specific hunting, fishing, and shooting equipment. During a twelve-year period (2000–2012) in the United States, $308 million from Federal Aid in Wildlife Restoration Funds were available to states for the acquisition of more than 74 million acres of wildlife habitat.[27] In addition, such wildlife conservation organizations as the Rocky Mountain Elk Foundation, Wildlife Habitat Canada, the Mule Deer Foundation, the Nature Conservancy, Ducks Unlimited, the Canadian Wildlife Federation, the National Wild Turkey Federation, Pheasants Forever, the Wild Sheep Foundation, and myriad others used private donations to purchase land or conservation easements on large tracts of wildlife habitat. Most of these areas are purchased with game animals in mind, but the wetlands acquired for waterfowl, forests purchased for deer or turkeys, mountainous areas protected for wild sheep, and grasslands restored for quail and pronghorn have also benefited nongame and endangered species that rely on those habitats. Recent estimates indicate about 70 percent of the users of these areas do not hunt, and in some properties that percentage may be as high as 95 percent.[28] Despite the funding source, conflicts sometimes arise when nonconsumptive users express concern about seeing hunters using such areas.

Funding sources for habitat vary among Canadian provinces. Some provinces have an association or trust fund that is dedicated to land acquisition. For example, in British Columbia, surcharges collected on hunting, angling, trapping, and guide-outfitting licenses are received by a trust fund managed by the Habitat Conservation Trust Foundation (HCTF), an independent organization. The trust fund is also sometimes the beneficiary of monetary court awards arising from the prosecution of environmental infractions. HCTF annually invests \$5–\$6 million in habitat acquisition, restoration, and enhancement, as well as other priority conservation projects throughout the province.

## Population Restoration

The restoration of wildlife populations across North America is arguably one of the greatest wildlife success stories in the history of modern wildlife conservation. Indeed, conservation efforts, frequently by hunters, had begun to restore overexploited populations even before the full development and implementation of the North American Model. Species like Canada geese (*Branta canadensis*), wood duck (*Aix sponsa*), white-tailed deer, pronghorn (*Antilocapra americana*), bighorn sheep (*Ovis canadensis*), and wild turkeys (*Meleagris gallopavo*) have benefited from restoration efforts driven by hunters concerned for the aesthetic and utilitarian loss represented by their demise.

The state of Arizona began restoring desert bighorn sheep in 1955 with reintroductions to historical habitat. Since then, more than one hundred translocations of at least 1,800 bighorn sheep have restored this iconic species in all previously occupied habitat and many other areas of suitable habitat.[29] Across North America, wild sheep (*Ovis* spp.) populations have been restored with more than 1,500 translocations involving about 25,000 wild sheep since 1922, exemplifying the type of wildlife restoration efforts that have occurred for decades in many states and provinces in North America with the funds generated from the regulated harvest of a few species.[30]

Compared to other continents, North America now has nearly a full complement of its native wildlife, despite habitats having been fractured and significantly challenged in some regions over the last three hundred years. Restoration of large mammal populations continues today, with elk (*Cervus canadensis*) being successfully translocated into historical ranges in the East. Work also continues for other species, such as bison (*Bison bison*), and some large predators, though with the restoration success of hunted species, focus has shifted to restoring nonhunted species, particularly threatened and endangered animals.

## Monitoring Wildlife Populations

Wildlife population monitoring and the accumulation of baseline trend data are fundamental to well-informed, science-based decision-making, a foundation of the North American model.[31] Though hunted species are not the only species monitored, they are typically a major focus. State, provincial, and federal agencies have a long history of monitoring wildlife populations, beginning at the very genesis of wildlife conservation in North America. Many agencies employ monitoring programs that have remained relatively consistent for decades and provide valuable trend data. The Canadian Wildlife Service (CWS) and the US Fish and Wildlife Service (USFWS), for instance, have conducted continent-wide aerial waterfowl surveys since 1955. This cooperative survey effort involves flying nearly 80,000 miles of survey each year throughout waterfowl areas from southern Mexico to northern Canada. Monitoring efforts by state and provincial wildlife agencies are now supplemented, and sometimes assisted, by a range of nongovernmental, hunting-based conservation organizations.

## Research

The North American model prescribes that management decisions be based in science. In the United States, about $57 million was apportioned in 2009 to state wildlife agencies from the Federal Aid in Wildlife Restoration Program for conducting more than 10,000 wildlife research projects.[32] The need for applied research to learn about problems affecting wildlife and the need for trained biologists to staff management agencies were both identified in Aldo Leopold's 1930 American Game Policy.

The identification of these needs led to the establishment in 1935 in the United States of a national system of Cooperative Wildlife Research Units, modeled after the first unit created in 1932 in Iowa. Each unit is cooperatively run by the federal Department of the Interior, the state fish and wildlife agency, the state land grant university, and the Wildlife Management Institute. The mission of each unit, now totaling forty nationwide and including fisheries research, is to meet the actionable science needs of its cooperators, to train agency practitioners in how to apply and interpret new science developments, and to develop the future workforce through graduate education and cooperative mentoring. Research funding comes from a variety of sources, but a significant portion of that funding has come from hunters and anglers through their state hunting and fishing license revenues and Wildlife and Sport Fish Restoration excise taxes on hunting, fishing, and shooting equipment purchases.

Canadian wildlife researchers obtain funding from a variety of sources, including but not limited to provincial agencies. For example, in British Columbia, the HCTF allocates funding for research. Wildlife conservation organizations (e.g., the Rocky Mountain Elk Foundation, the Alberta Conservation Association, the WSF) and extractive industries (e.g., energy companies and forestry) also contribute money for wildlife research.

## Law Enforcement

One of the most important wildlife conservation expenditures directly supported by hunters and other consumptive users is law enforcement. Currently, thousands of wildlife conservation law enforcement officers are actively working in the United States and Canada, and most are paid with income from the sales of hunting, trapping, and fishing licenses. Though wildlife law enforcement officers are dedicated to protecting natural resources, they also perform duties related to wildlife monitoring, water quality, habitat protection, public safety, search and rescue, littering, vandalism, and trade in threatened and endangered species. Additionally, they provide backup to other local law enforcement agencies.

Efforts to uphold the complex array of legal restrictions to wildlife harvest are essential, and they constitute an obvious example of the community taking care of that in which its members have a personal vested interest. Regulated hunting remains regulated so long as laws remain enforced. In addition to providing financial support, surveys consistently report that about 50 percent of hunters and anglers have had recent contact with enforcement personnel in the field and hold them in high regard.[33]

## Agency Infrastructure

In recent decades, some state and provincial wildlife agencies have become creative in their ability to garner additional funding sources to supplement the long-standing contribution of hunters, anglers, trappers, and shooting enthusiasts. Funds from lotteries, income tax, special stamps, and the like are sometimes channeled to wildlife agencies and earmarked for programs that have not received adequate financial support in the past (e.g., nongame or habitat acquisition and management). These recent supplemental funds are an important addition to the budgets of wildlife agencies, but they are vulnerable to legislative pressures and do not replace or negate the importance of the base funds derived from consumptive activities. Supplemental funds are effective because there exists an agency infrastructure able to apply them directly to specific program areas. Traditional funds derived from consumptive wildlife use remain the primary support for law enforcement, personnel resources, and all other day-to-day agency operations. The agencies most effective in conserving all wild resources are those that maintain a solid financial foundation, provided by monies derived from the well-regulated harvest of select wildlife species.

## Conservation of Nonhunted Species

In North America, a preponderance of hunter-generated money is still expended on the management and protection of hunted species. In order to ensure long-term sustainability, populations of hunted species require a greater intensity of monitoring, law enforcement, management, and research.

Conservation programs for nongame species in the United States are funded through a variety of sources, including income tax write-offs, special stamps, independent grants, donations, some lottery or other gambling revenue, and some sales tax. A portion of hunters' dollars from the Federal Aid in Wildlife Restoration Program is also earmarked for the conservation of nonhunted species, and this funding mechanism generates millions of dollars annually. Through the current authorization of the Wildlife Restoration Program, hunters' dollars contribute to the restoration of many threatened and endangered species, such as the California condor (*Gymnogyps californianus*), the Mexican wolf (*Canis lupus baileyi*), the black-tailed prairie dog (*Cynomys ludovicianus*), and the black-footed ferret (*Mustela nigripes*). Under this system, conservation actions also have the potential to protect wildlife species before they become threatened or endangered.[34] However, this is a small percentage of what is needed and does not begin to fully address the needs of taxonomic groups including native fish, songbirds, amphibians, and reptiles. As a consequence, state wildlife agencies are challenged to find and maintain additional funds for management of species that were not heavily exploited historically, and, therefore, are not the major focus of the Wildlife Restoration Program. This has been identified by some as a weakness of the North American Model of Wildlife Conservation. The future of conservation in North America will have to not only include but also extend beyond the existing model of using hunters' dollars to conserve wildlife diversity.

## Information and Public Relations

Communicating with the public and considering human dimensions in wildlife management are vital to the effectiveness of wildlife agencies.[35] Most such agencies now have public information officers on staff to disseminate wildlife information and to update stakeholders on agency activities through press releases, websites, social media, radio, television, and a multitude of publications for diverse audiences. Some US wildlife agencies use Federal Aid in Wildlife Restoration Funds for this purpose, but most use money garnered from the sale of hunting and fishing licenses. In this way, the wider public benefits from the baseline funding provided by regulated hunting. In Canada, much of the funding for these kinds of activities comes from taxes, which sometimes poses a problem for wildlife agencies as funding cuts and downsizing can erode their outreach programs and communications efforts.

## Advocacy for Wildlife and Wild Lands

Management guidelines establish limits to hunting and trapping activities, and law enforcement officers enforce regulations. Much of North America's conservation success, however, is due to incentives based on self-interest and personal ethics.[36] Many outdoorspeople in North America practice what may be perceived as even "stronger personal ethics" than the law requires.[37] Historically, hunting has been a primary force in assuring wildlife a place on the landscape. Having a vested interest in wildlife abundance fosters attention, and wildlife thrives with attention, while it withers from neglect.[38] The strong desire to hunt wildlife appears to be resistant to social change. Frequently, the modern passion to hunt expresses itself as a deep, lifelong interest in and devotion to wildlife, often accompanied by considerable practical work, even sacrifice, by the hunter on behalf of wildlife. Witness the many organizations dedicated to the conservation of wildlife in North America. Arizona's first desert bighorn sheep hunt occurred in 1953; two years later the state wildlife agency began its aggressive translocation program, which has now restored all of the state's historical wild sheep populations. There are many examples of hunter advocacy being instrumental in implementing game management and wildlife conservation on a broad scale.[39] Aldo Leopold, considered the founder of the wildlife management profession, was an avid and lifelong hunter.[40]

## Political Support

Early groups of organized hunters were instrumental in providing the political support necessary to implement many of the laws that coalesced into the system of conservation we have today. For example, Theodore Roosevelt organized the Boone and Crockett Club in 1887 by assembling some of the most powerful and influential conservation-minded people of the day, many of them hunters. The Boone and Crockett Club successfully lobbied for the establishment of Yellowstone National Park, preservation of the bison, cessation of market hunting, and improved wildlife enforcement, as well as the establishment of a continental treaty for waterfowl.

A connection between politics and conservation is inescapable. When political influence is at odds with wildlife conservation efforts, sportsmen and women at local and national levels have traditionally demonstrated their willingness and ability to unite in support of wildlife. Though nonhunters are also often involved, it is frequently the organizational infrastructure of hunting organizations that drives coordinated efforts to influence policymakers and advocate for wildlife. This same infrastructure is also frequently used for mounting campaigns in defense of crucial wildlife habitat threatened by conflicting interests. With recent declines in hunter retention and recruitment, effective wildlife advocacy may also decline, or become less vocal. As wildlife and habitat are subjected to increasing pressures, hunters and nonhunters will need to focus on common goals and combine their collective resources and efforts to ensure wildlife's continued existence.

## Biological Samples and Information

Hunters are an important source of biological information for wildlife managers. Data such as the total number of animals harvested, sex and age ratios, body weight or condition, and harvest location have been collected at hunter check stations since the early years of wildlife management. Other hunt-related information is also routinely collected at check stations, in the field, or with post-hunt questionnaires by phone or mail. Whether the hunter was successful, the total number of hunter days expended, the area hunted, and other information can be used to track trends in wildlife population parameters or abundance. In some cases, these check stations and questionnaires are mandatory, but in many instances hunters voluntarily provide information that might help managers.

Biological samples from harvested animals are used to determine prior disease exposure, parasite loads, nutritional status, genetic relationships or diversity, body size, and age. The collection of these types of samples is sometimes done by the hunters themselves and requires cooperation and commitment. Along with other members of the public, hunters routinely provide information on species distribution and sources of unusual mortality. Such input is valuable in tracking changes in wildlife distributions and health in the face of habitat loss, natural disasters, climate change, and emerging diseases.

### Volunteerism

Individual hunters and trappers, along with the organizations to which they belong, have remained active in providing volunteer labor for habitat improvement projects, constructing nesting structures or boxes, altering fences to be wildlife friendly, teaching hunting and trapping education courses, conducting wildlife surveys, working wildlife check stations, conducting routine facility maintenance, cleaning up trash, and many other valuable activities.[41] These volunteer efforts benefit wildlife directly and allow wildlife management agencies to stretch their conservation dollars to accomplish additional goals.

Most of these projects benefit both hunted and nonhunted species. For example, big-game hunters throughout the US West have installed water collection and retention devices for specific game species, such as wild sheep, but countless numbers of bird, mammal, insect, reptile, and amphibian species use them, too. Designs of water catchments for large mammals have been altered through the years to accommodate the needs of bats and smaller terrestrial nongame animals.

Residents in the state of Maryland were asked if they would be interested in volunteering their time with the state wildlife agency.[42] Results revealed that 22 percent of hunters were "very likely" to volunteer, as compared to 7 percent of the nonhunters. This helps illustrate the level of commitment to stewardship of natural resources that is inherent in the hunting community

### Foundational Changes in the Human-Wildlife Relationship and a Changing Context in North America

In North America today, the number of hunters is decreasing, while the overall human population is growing.[43] From 1980 to 1996, the percentage of Americans who hunted dropped from 10 percent to 5 percent.[44] Meanwhile, the public's perceptions of and interactions with nature have changed tremendously in recent decades. Vast numbers of people now reside in urban settings; they no longer grow, forage, hunt, or butcher their own food, but rely instead on supermarkets and restaurants. For many individuals, firsthand ties to nature no longer exist. This has had a profound impact on how modern humans perceive the natural world. It has also influenced individual and social attitudes about specific wildlife management actions, including regulated hunting, which is perceived by some individuals as a cruel and unnecessary anachronism, rather than as a tool for wildlife management and conservation.

An increasing number of people obtain their information about nature from television and the Internet. Digital presentations of nature often focus on individual animals and not on the realities of managing populations and executing successful conservation programs. Similarly, such displays of nature, created primarily for the purpose of human entertainment, may not always be realistic and, in some cases, may create distorted or inaccurate views of traditional wildlife management activities, including hunting. This may lead to diminished public support for the consumptive activities that support conservation efforts.

### Everyone Has a Vested Interest in Conservation

Though hunters played a lead role in the development of the North American Model, this does not mean they own the future of this paradigm. Social changes are already challenging the foundation of conservation through consumption. Those who wish to perpetuate a realistic and proven conservation model will need to work to preserve it. Future efforts to conserve wildlife and wild places will not succeed without a broad base of public support.[45] History has illustrated the failure of conservation prescribed by the elite and fashioned after the desires of the minority. Stewardship of wildlife has always required much hard work and that will not change; nor will the need for advocacy backed by science-based research, education, and management programs.

Today in North America, regulated hunting does still enjoy wide-ranging public support. Several surveys have consistently reported that 75–81 percent of respondents support hunting and agree it should continue.[46] Trend data from sequential surveys indicate there may be an increasing proportion of Americans who approve of legal fair-chase hunting when the harvest is utilized.[47] However, continued broad public support for hunting and other consumptive or sustainable use activities is not guaranteed and must not be taken for granted.

The continued success of hunters, trappers, and anglers in supporting conservation will depend on a public that understands how the current system works. Many individuals enjoy the outdoors. A camping, hiking, or hunting trip is typically much more enjoyable when wildlife is seen. And yet, many people do not understand what drives and funds the programs that provide the foundation for the nature-based opportunities they enjoy. The founders of the North American Model worked hard to turn the tide of public opinion in support of a widespread conservation ethic.[48] Those interested in wildlife today must continue to work to increase public awareness of this uniquely effective system. The future success of the North American Model will require hunters and trappers to remain relevant to conservation.[49] Hunters should be recognized for their past, present, and potential future contributions and should be viewed as supporting wildlife conservation efforts.[50]

## Conclusion

All people benefit from wildlife; everyone has a vested interest in the perpetuation of abundant wildlife resources. The North American Model of Wildlife Conservation, which integrates hunting with conservation practices, has been recognized internationally as a successful approach under specific conditions, which offers realistic long-term solutions to wildlife depletion and landscape impoverishment.[51] Today, this continental approach to conservation through sustainable use continues to rely heavily on economic, political, and social contributions by hunters.

Moving forward, consumptive wildlife users must be more cognizant of public perception. Hunters must work to dispel the negative imagery and headlines that have recently plagued the hunting community and eroded some of the public's confidence in a hunter-based system of conservation. Hunters and other consumptive users must continue to demonstrate and articulate that they are truly stewards of all wildlife, and are not concerned only with game species.[52] How the hunter communicates and positions himself or herself in the minds of the nonhunting public will ultimately decide whether hunting will continue far into the future, and this will directly impact, and perhaps determine, the future of the North American Model.

*NOTES*

1. S. P. Mahoney and J. J. Jackson, "Enshrining Hunting as a Foundation for Conservation—The North American Model," *International Journal of Environmental Studies* 70, no. 3 (2013): 448–459.
2. J. R. Heffelfinger, V. Geist, and W. Wishart, "The Role of Hunting in North American Wildlife Conservation," *International Journal of Environmental Studies* 70, no. 3 (2013): 399–413; Mahoney and Jackson, "Enshrining Hunting."
3. Mahoney and Jackson, "Enshrining Hunting."
4. T. Roosevelt, T. S. Van Dyke, D. G. Eliot, and A. J. Stone, *The Deer Family* (New York: Macmillan, 1902); G. B. Grinnell, "The Game Preservation Committee," in *Hunting at High Altitudes*, ed. G. B. Grinnell (New York: Harper and Brothers, 1913), 421–432; A. Leopold, *Game Management* (New York: Charles Scribner's Sons,

1933); J. F. Reiger, *American Sportsmen and the Origins of Conservation*, rev. ed. (Norman: University of Oklahoma Press, 1986); N. L. Bradley, "How Hunting Affected Aldo Leopold's Thinking and His Commitment to a Land Ethic," *Proceedings of the Fourth Annual Governor's Symposium on North American Hunting*, hosted by the North American Hunting Club, Minnetonka Minnesota (1995): 10–13; E. D. Thomas Jr., *How Sportsmen Saved the World: The Unsung Conservation Efforts of Hunters and Anglers* (Guilford, CT: The Lyons Press, 2009).

5. V. Geist, "The North American Model of Wildlife Conservation: A Means of Creating Wealth and Protecting Public Health while Generating Biodiversity," in *Gaining Ground: In Pursuit of Ecological Sustainability*, ed. D. M. Lavigne (Guelph, Ontario: International Fund for Animal Welfare, 2006), 285–293; R. Patterson, "Executive Summary: The Symposium on the Ecologic and Economic Benefits of Hunting," in *World Symposium: Ecologic and Economic Benefits of Hunting, Proceedings of the Symposium of Hunting Activities* (Windhoek, Namibia: World Forum on the Future of Sport Shooting Activities, 2010), 390–391; M. D. Duda, M. F. Jones, A. Criscione, and A. Ritchie, "The Importance of Hunting and the Shooting Sports on State, National and Global Economies," in *World Symposium: Ecologic and Economic Benefits of Hunting, Proceedings of the Symposium of Hunting Activities* (Windhoek, Namibia: World Forum on the Future of Sport Shooting Activities, 2010), 276–293.
6. Geist, "The North American Model of Wildlife Conservation"; Heffelfinger, Geist, and Wishart, "The Role of Hunting"; Mahoney and Jackson, "Enshrining Hunting."
7. A. Leopold, "Wildlife in American Culture," *Journal of Wildlife Management* 7, no. 1 (1943): 1–6.
8. R. Southwick and T. Allen, "Expenditures, Economic Impacts and Conservation Contributions of Hunters in the United States," in *World Symposium: Ecologic and Economic Benefits of Hunting, Proceedings of the Symposium of Hunting Activities* (Windhoek, Namibia: World Forum on the Future of Sport Shooting Activities, 2010), 308–313.
9. S. P. Mahoney, "Recreational Hunting and Sustainable Wildlife Use in North America," in *Recreational Hunting, Conservation and Rural Livelihoods*, ed. B. Dickson, J. Hutton, and W. M. Adams (Oxford, UK: Wiley-Blackwell, 2009), 266–281.
10. J. Baughman and M. King, "Funding the North American Model of Wildlife Conservation in the United States," in *Strengthening America's Hunting Heritage and Wildlife Conservation in the 21st Century: Challenges and Opportunities*, ed. J. Nobile and M. D. Duda (Harrisonburg, VA: Sporting Conservation Council, 2008), 57–64.
11. Mahoney and Jackson, "Enshrining Hunting."
12. D. J. Decker, T. L. Brown, and W. F. Siemer, *Human Dimensions in Wildlife Management in North America* (Bethesda, MD: The Wildlife Society, 2001); M. D. Duda, M. F. Jones, and A. Criscione, *The Sportsman's Voice: Hunting and Fishing in America* (State College, PA: Venture Publishing, 2010).
13. Heffelfinger, Geist, and Wishart, "The Role of Hunting."
14. C. D. Meine, *Aldo Leopold: His Life and Work* (Madison: University of Wisconsin Press, 1991).
15. M. R. Conover, "Effect of Hunting and Trapping on Wildlife Damage," *Wildlife Society Bulletin* 29, no. 2 (2001): 521–532.
16. L. H. Carpenter, "Harvest Management Goals," in *Ecology and Management of Large Mammals in North America*, ed. S. Demarais and P. R. Krausman (Upper Saddle River, NJ: Pearson, 2000), 192–213.
17. M. R. Conover, *Resolving Human-Wildlife Conflicts: The Science of Wildlife Damage Management* (Boca Raton, FL: CRC Press, 2002); Duda et al., "The Importance of Hunting."
18. M. R. Conover, W. C. Pitt, K. K. Kessler, T. J. DuBow, and W. A. Sanborn, "Review of Human Injuries, Illnesses, and Economic Losses Caused by Wildlife in the United States," *Wildlife Society Bulletin* 23, no. 3 (1995): 407–414.
19. T. L. Brown, D. J. Decker, S. J. Riley, J. W. Enck, T. B. Lauber, P. D. Curtis, and G. F. Mattfeld, "The Future of Hunting as a Mechanism to Control White-Tailed Deer Populations," *Wildlife Society Bulletin* 28, no. 4 (2000): 797–807.
20. M. D. Duda, M. F. Jones, and K. C. Young, *Wildlife and the American Mind: Public Opinion on and Attitudes toward Fish and Wildlife Management* (Harrisonburg, VA: Responsive Management, 1998).
21. US Fish and Wildlife Service, "Final Apportionment of Pittman-Robertson Wildlife Restoration Funds (CFDA #15.611) for Fiscal Year 2016," February 22, 2016, https://wsfrprograms.fws.gov/Subpages/GrantPrograms/WR/WRFinalApportionment2016.pdf.
22. Southwick and Allen, "Expenditures," 308–313.
23. E. B. Arnett and R. Southwick, "Economic and Social Benefits of Hunting in North America," *International Journal of Environmental Studies* 72, no. 5 (2015): 734–745.
24. K. Hurley, C. Brewer, and G. Thornton, "The Role of Hunters in Conservation, Restoration, and Management of North American Wild Sheep," *International Journal of Environmental Studies* 72, no. 5 (2015):

784–796; R. Lee, former Chief Executive Officer, Wild Sheep Foundation, conversation with author, 2016.
25. Rocky Mountain Elk Foundation, "RMEF Mission Statement & Fast Facts," April 12, 2012, http://www.rmef.org/NewsandMedia/PressRoom/AboutRMEF/MissionFastFacts.aspx.
26. T. W. Hughes and K. Lee, "The Role of Recreational Hunting in the Recovery and Conservation of the Wild Turkey (*Meleagris gallopavo* spp.) in North America," *International Journal of Environmental Studies* 72, no. 5 (2015): 797–809; J. E. Kennamer, Chief Conservation Officer, National Wild Turkey Federation, conversation with author, 2016.
27. Debbie Wircenske, Grant Management Specialist, US Fish and Wildlife Service, conversation with author, 2016.
28. US Fish and Wildlife Service, "Pittman-Robertson Wildlife Restoration, 2013 Budget Justification," December 20, 2013, https://www.fws.gov/budget/2013/PDF%20Files%20FY%202013%20Greenbook/24.%20Wildlife%20Restoration.pdf.
29. J. O'Dell, "50 Years and a Lot of Sheep Later," *Arizona Wildlife Views*, November–December 2007, 8–12.
30. Kevin Hurley, Vice President of Conservation and Operations, Wild Sheep Foundation, conversation with author, 2016; Hurley, Brewer, and Thornton, "The Role of Hunters."
31. Baughman and King, "Funding the North American Model," 57–64.
32. Debbie Wircenske, personal communication.
33. Duda, Bissell, and Young, *Wildlife and the American Mind.*
34. Baughman and King, "Funding the North American Model."
35. Decker, Brown, and Siemer, *Human Dimensions.*
36. Leopold, *Game Management.*
37. Heffelfinger, Geist, and Wishart, "The Role of Hunting."
38. Geist, "The North American Model."
39. Mahoney and Jackson, "Enshrining Hunting."
40. Bradley, "Aldo Leopold's Thinking"; R. B. Peyton, "Wildlife Management: Cropping to Manage or Managing to Crop?" *Wildlife Society Bulletin* 28, no. 4 (2000): 774–779.
41. V. C. Bleich, "Affiliations of Volunteers Participating in California Wildlife Water Development Projects," in *Wildlife Water Development*, ed. G. K. Tsukamoto and S. J. Stiver (Reno: Nevada Department of Wildlife, 1990), 187–192.
42. Responsive Management, *Wildlife Viewing in Maryland: Participation, Opinions and Attitudes of Adult Maryland Residents towards a Watchable Wildlife Program*, report prepared for Maryland Wildlife Division (Harrisonburg, VA: Responsive Management, 1993).
43. Duda et al., "The Importance of Hunting."
44. M. A. Schuett, D. Scott, and J. O'Leary, "Social and Demographic Trends Affecting Fish and Wildlife Management," in *Wildlife and Society: The Science of Human Dimensions*, ed. M. J. Manfredo, J. J. Vaske, P. J. Brown, D. J. Decker, and E. A. Duke (Washington, DC: Island Press, 2009), 18–30.
45. S. P. Mahoney, "The Importance of How Society Views Hunting," *Proceedings of the 87th Annual Conference of the Western Association of Fish and Wildlife Agencies* (2007): 62–67.
46. Duda, Jones, and Criscione, *Sportsman's Voice*, 32.
47. Duda, Jones, and Criscione, *Sportsman's Voice*, 44.
48. Reiger, *American Sportsmen*; Mahoney, "How Society Views Hunting."
49. Mahoney, "How Society Views Hunting."
50. V. Geist, S. P. Mahoney, and J. F. Organ, "Why Hunting Has Defined the North American Model of Wildlife Conservation," *Transactions of the North American Wildlife and Natural Resources Conference* 66 (2001): 175–185.
51. Mahoney and Jackson, "Enshrining Hunting."
52. J. Posewitz, *Beyond Fair Chase: The Ethic and Tradition of Hunting* (Guilford, CT: Falcon, 2002).

# Science and the North American Model

## *Edifice of Knowledge, Exemplar for Conservation*

James A. Schaefer

*Science is an underpinning of the North American Model. As such, it is the basis for informed decision-making in wildlife management in Canada and the United States. While the history of wildlife management is deeply rooted in the writings of Aldo Leopold, it is often the communication of this knowledge to wildlife "users" that garners the most support for conservation and sustainable use of wildlife resources. This chapter discusses the historical development of scientific management of wildlife and examines how the relatively new field of "human dimensions" has evolved the theory and application of scientific management to include the influence of humans in the understanding of wildlife and habitats. This chapter also comments on hunters' contributions to the scientific management of wildlife species and provides recommendations for scientific advocacy, emphasizing the need for scholarship in transferring science to wildlife conservation and management policy.*

### Edifice Rising

Science is powerful and precious. It generates evidence—the factually verifiable accounts of the natural world capable of supporting public policy.[1] It engenders trust—the faith of citizens that science will improve the quality of their lives. In North America, the public regards scientists as esteemed professionals, their prestige on par with medical practitioners and Supreme Court judges.[2] Science lays claim to the most successful human enterprise, ever.[3] It is a pinnacle of human achievement.

The triumph of the North American Model of Wildlife Conservation can be traced, in large measure, to evidence-based policies and public trust, engendered by science. Today, wildlife science radiates accomplishment. Take, for example, the accumulated scholarly articles on wildlife published since 1991 (the year of my first paper). If printed and stacked, they would reach an imposing height of nearly ten stories. In 2017, nearly four thousand papers were added to this edifice; its height is doubling every six years. The rise in society memberships, conferences, journal subscriptions, and academic programs reinforce the towering impression of conservation science as being more successful than ever.[4]

To the career academic, this edifice is an obsession. My students, colleagues, and I typically focus on stylish developments on the top floor, often disregarding the foundation. That sturdy base was laid by the firm hands and sharp minds of early North American sportsmen, motivated by decimation and disappearance of species in the late nineteenth century. The conservation model arose "phoenix-like from the ashes of our dead wildlife."[5] Sportsmen remain some of its key builders—the proponents and providers of field observations, samples, funding, lobbying, and questions that bolster scientific inquiry.

The successes of modern wildlife science are irrefutable. Its loftiness, however, does not guarantee effectiveness.[6] To make real contributions—to be considered in people's decisions and actions—demands a structure that is valued, coherent, and open to the public.[7]

This shortcoming of science has long been recognized. In the 1960s, John Platt observed researchers' satisfaction merely to toss another brick into an incoherent pile of information, noting that "most such bricks just lie around the brickyard."[8] At times, ecology seems adrift in data, yet deficient in knowledge. The journals, remarked Paul Ehrlich, have become "clogged with dribs and drabs of information on a vast variety of organisms and communities—increasingly sophisticated studies of more and more trivial problems."[9]

Harsh words, indeed. But to an aspiring scholar, under the pressure of "publish-or-pogey," bricks are the currency that generates academic rewards: hiring, funding, tenure, and promotion. So, the solitude of the ivory tower endures. Since Platt, at least a generation of scientists has come and gone. Andrew Hoffman has witnessed little change: "We find ourselves talking to smaller and narrower academic audiences, using a language that an educated reader does not understand, publishing in journals they don't read and asking questions for which they have little concern. Whether this work actually creates real world change is a question that is rarely, if ever, asked."[10] The mortar of public engagement—what makes science relevant—is regarded as a distraction.

## Worlds Apart

Science may stand as a pillar of the North American Model, but evidence-based professional practice can be surprisingly far from the norm.[11] Wildlife managers seldom rely on wildlife science, and instead draw on conventional wisdom and common sense. This custom appears to be widespread in North America and elsewhere. Some examples:

- In a widely cited study of an English wetland, just 2 percent of conservation actions were grounded in verifiable evidence; 77 percent were solely experience-based.[12]
- In 185 conservation plans for US National Wildlife Refuges, 96 percent failed to indicate any follow-up actions from monitoring; only one plan dared use the "h" word—hypothesis.[13]
- In performance assessments of two Australian conservation agencies, whose responsibility included more than one thousand protected areas, experience-based knowledge was used five times more frequently than evidence-based knowledge; 90 percent of assessments were made without evidence-based knowledge.[14]
- Most US resource management agencies have avoided adaptive management (AM) in favor of "AM lite."[15] They opt for ad hoc contingency planning over the deliberate testing of management activities—that is, by confronting hypotheses with data.

If science is not routinely put to use by wildlife professionals, its disconnect from citizens becomes even greater. Roughly half of North Americans do not have a grasp of elementary scientific facts—for instance, that it takes one year for the earth to orbit the sun or that antibiotics do not kill viruses.[16] Not surprisingly, scientists and the public may find themselves poles apart on complex and crucial matters of environment and health (see table 8.1). Consider evolution: whereas biologists recognize that "nothing makes sense" without evolution, its acceptance among US adults is roughly 40 percent, and in my home province of Ontario, just 51 percent of people appreciate that "human beings evolved from less advanced forms over millions of years."[17] Evolution by natural selection—the linchpin of biological understanding—fails to resonate with millions of North Americans, even though it is pivotal to mod-

*Table 8.1.* Worlds apart: levels of acceptance by scientists and the public of contemporary environmental issues

| Statement | Acceptance by Scientists | Acceptance by Public | Reference |
|---|---|---|---|
| Natural selection plays a role in evolution. | 87% | 32% | Funk & Rainie (2015)[a] |
| Genetically modified foods are safe to eat. | 88% | 37% | Funk & Rainie (2015)[b] |
| There is solid evidence temperatures on earth have increased during the past four decades. | 97% | 65% | Lachapelle, Borick, & Rabe (2012)[c] |

*Source*: Adapted from A. J. Hoffman, "Reflections: Academia's Emerging Crisis of Relevance and the Consequent Role of the Engaged Scholar," *Journal of Change Management* 16, no. 2 (2016): 77–96.

[a] C. Funk and L. Rainie, "Public and Scientists' Views on Science and Society," Pew Research Center, January 29, 2015, http://assets.pewresearch.org/wp-content/uploads/sites/14/2015/01/PI_ScienceandSociety_Report_012915.pdf.

[b] Funk and Rainie, "Public and Scientists' Views on Science and Society."

[c] E. Lachappelle, C. P. Borick, and B. G. Rabe, "Public Attitudes toward Climate Science and Climate Policy in Federal Systems: Canada and the United States Compared," *Review of Policy Research* 29, no. 3 (2012): 334–357.

ern matters of health and wealth: why antibiotics lose their potency, why DDT fails to eliminate malaria-carrying mosquitoes, why trophy hunting can leave fewer trophy animals, and why commercial fishing can reduce the size of fish.[18]

This is no mere inconvenience. The rift between scientists and citizens carries immense consequences for people and wildlife, particularly as we surge headlong into the sixth mass extinction.[19] Public ignorance of science will reverberate in the forests, prairies, lakes, mangroves, and oceans where the fate of countless species is being determined. Conservation makes a demonstrable difference in averting extinction, but achieving those outcomes demands purposeful steps by scientists to make their science accessible and pertinent.[20]

This challenge is becoming steeper. In a society more urbanized and more distant from nature, the obstacles to effective conservation science are growing. A glimpse of this future landscape can be found in everyday childhood experiences. A growing proportion of children, for instance, has never climbed a tree.[21] The *Oxford Junior English Dictionary* recently dropped one hundred words linked to nature—nouns such as "acorn," "beaver," "chestnut," "cygnet," "minnow," and "raven."[22] If conservationist Papa Dioum is correct—that we conserve only what we love, love only what we understand, and understand only what we have been taught—the playgrounds of today foretell the wildlife of tomorrow at risk: unknown, unappreciated, and uncared for.

## An Underpinning with a History

The North American Model is not just a paradigm of successful conservation. At its finest, it is an exemplar of engagement between scientists and users of scientific knowledge—a paradigm of effective science. Its achievements—the habitats and wildlife populations restored, the areas protected, and the species rescued from the brink—are numerous and rising.[23] Conservation is a long-term business.[24] These outcomes are typically the result of decades of effort, grounded in traditions more than a century old.

History is important; it serves as "a vaccine against bad ideas."[25] Yet society can easily fall into collective amnesia, forgetful of events a mere handful of generations past. As anthropologist Wade Davis remarked, for most North Americans, the time of the buffalo "is as distant from their lives as the fall of Rome." Understanding the past is key to our environmental future.

Hence, as Davis observed, "This capacity to forget, this fluidity of memory, has dire implications in a world dense with people."[26] Even recent descriptions of conservation biology as "new" are not new. In the first sentence of the first article of the first volume of the *Journal of Wildlife Management*, more than eighty years ago, you can read about the "new and growing field of conservation biology."[27]

The roots of the North American Model can be found in the decimation of passenger pigeons, Labrador ducks, plains bison, muskoxen, elephant seals, and great auks. By the late 1800s, it had become obvious that renewable resources were "far from illimitable."[28] The collapse of the bison, in particular, transformed the continental interior dominated by perhaps 30 million beasts in the mid-eighteenth century to just a few hundred by the early twentieth century—a brush with extinction effected in just one and a half decades. The decimation was swift and obvious. Andrew Isenberg recounted a view of the prairie landscape in the 1880s, as seen by an American rancher, who found himself "never out of sight of a dead buffalo and never in sight of a live one."[29]

The focus of early conservationists was to safeguard living resources by establishing parks, implementing harvest regulations, and hiring game wardens. The emphasis was on wise use—a parallel with the "intelligent tinkering" of Aldo Leopold and "wise economy" of George Perkins Marsh.[30] The goal, in Clifford Sifton's words, was to provide "the best and most highly economic development and exploitation in the interests of the people."[31] Science was part of the initiative. The founding fathers of North American conservation espoused science; it was the means to fill the great knowledge gaps about plants and animals on the continent.[32]

This was a response to crisis. In Ontario, for instance, the nineteenth-century decline and disappearance of fish and game—including species only recently recovered, such as wapiti and wild turkey—spurred the provincial government to establish a Royal Commission on Fish and Game. Its mandate was to review the status of wildlife; its report was submitted in 1892. As Janet Foster noted, public opinion and representations by sportsmen compelled the commissioners to acknowledge that, without prompt action, Ontario's wildlife would be devastated and "what had been for years a sportsmen's paradise, known all over the world, and furnishing game and fish in abundance . . . would be no more."[33] In the wake of the commission's report, the provincial government broadened and bolstered wildlife acts, set up a new Game and Fisheries Board, and, by the end of 1892, had hired 392 deputy game wardens. The provincial parks of Algonquin (in 1893) and Rondeau (in 1894) were inaugurated as wildlife preserves.[34]

These echoes rebounded across the continent. Canadians were profoundly influenced by examples and experiences south of the border: the loss of the frontier, the role of civilization in declining wildlife, and the establishment and success of US national parks.[35] By the early twentieth century, the early formations of the North American Model were taking hold. President Theodore Roosevelt wrote to Prime Minister Wilfred Laurier—an invitation to Canadian representatives to attend a National Conservation Conference in Washington, February 1909. Months earlier, the Americans had held two conservation conferences: a Conference of Governors to confer with the president on matters of conservation, and a National Conservation Commission to complete inventories of natural resources. To Roosevelt, it was evident that resources were not confined by national boundaries and that there was a need for conserving them "as wide as the area upon which they exist."[36] The purpose of the conference was to consider matters of mutual concern.

Just three months later, on May 19, 1909, Parliament established the federal Commission for the Conservation of Natural Resources—a direct consequence of the recommendation in the Declaration of Principles.[37] The language was clear: "It shall be the duty of the Commission to take into consideration all questions which may be brought to its notice relating to the conservation and better utilization of the natural resources of Canada, to make such inventories, collect and disseminate such information, conduct such investigations, inside and outside of Canada."[38] Novel were both its mandate—independent, autonomous, objective, nonpartisan—and its composition—twelve ministers of the Crown, eight university professors, and twelve others. During its short lifetime, the commission was instrumental in developing the Migratory Birds Convention—"a treaty that has served ever since as a keystone of North American wildlife policy and protection"—as well as the National Parks Act, the Northwest Territories Game Act, and the Migratory Birds Convention Act. In the succeeding decades, science became intertwined with wildlife management at the state, provincial, and federal levels.[39] In this, Valerius Geist wrote, "North Americans were decades ahead of their time."[40]

In Canada, the commission was short lived (1909–1921), but the groundwork had been laid for a central role of science in conservation.[41] It was a signal, too, of broader societal change. The commission represented the new faith of citizens in disinterested scientific expertise and rationality.[42] That modernity of rationality endured through the twentieth century.

During the 1920s, public outreach by conservation scientists also become customary. In 1924, for example, four federal Canadian chief migratory bird officers tallied more than 452 public lectures.[43] Aldo Leopold emerged as an advocate for roadless wilderness and balanced conservation—his "land ethic." Leopold was a true force of nature; he united hunters, conservationists, and landowners with the idea and the goal of stewardship. Shortly after World War II, scientific involvement in conservation spread with the establishment of federal agencies: the Dominion (later Canadian) Wildlife Service and the US Fish and Wildlife Service. Cooperative Fish and Wildlife Research Units—unique partnerships among US universities and federal and state governments—proliferated to universities across thirty-eight states. In 1976, the US government created an office of chief science advisor and successive presidents have continued to appoint scientists as senior advisors—a tradition that can be traced to the pioneers of the North American Conservation Model: Audubon, Roosevelt, Grinnell, Pinchot, and Merriam.[44]

## An Uneasy Relationship

The lesson from history, philosopher Georg Wilhelm Friedrich Hegel wryly noted, is that we don't learn from history. And the story of the North American Model underscores the often thorny relationship between science and politics—a tension for which there are recent and historical examples. During the 1919 binational Conservation Conference, Canadian Interior Minister Arthur Meighen spoke directly of the importance of conservation of game and other natural resources. Actions, however, would soon speak more loudly. "Ironically," wrote J. Alexander Burnett, "Meighen, as a newly elected Conservative prime minister, would disband the Conservation Commission just two years later, citing its independence from departmental authority: 'I do not think it is consistent with our system of government that there should be a body for which no one is answerable and over which no one has any control.'"[45] Science, even if a pillar of the North American Model, resides uneasily and tenuously within government.

Despite the tension and divergent demands, government-supported wildlife science in North America has at times been able to find peaceful coexistence with politics. Burnett also noted that "many of the outstanding scientific and conservation achievements of the [Canadian] Wildlife Service over its first fifty years can reasonably be attributed to the maintenance of a healthy distance between field research and the political arena."[46]

But the strain—the independence of wildlife science versus the responsibility of politics—continues to emerge.[47] In Canada, during 2003–2006 for instance, the federal minister of the environment routinely rejected advice from the Committee on the Status of Endangered Wildlife in Canada, the independent scientific body responsible for evaluating species' status and for proposing those at risk for legal listing. Ministerial rejection has been inordinately frequent in the case of commercially harvested fishes and northern mammals—a bias against "fuzzy, tasty endangered species."[48] In 2008, the position of chief science advisor was rescinded. While the position was restored in 2017, following a change of government, Canada remains the only legislatively accountable body worldwide to ever have eliminated such a position.[49] And in 2013, reminiscent of the fate of the Conservation Commissions nearly a century earlier, the National Roundtable on the Economy and the Environment—an external body advising the federal government on matters of environmental sustainability and economic prosperity—was axed.

My colleagues are often dismayed by such decisions. Rather than chastise politicians and the public for their scientific ignorance, however, scientists might enhance their effectiveness by shedding their naïveté—by separating what is scientifically possible from what is culturally acceptable.[50] In an echo of Aldo Leopold decades earlier, Gregory van der Vink (1997) wrote: "In the next generation, we will need not only scientists who are experts in subspecialties, but also those with a broad understanding of science and a basic literacy in economics, international affairs, and policy-making. In the end, our greatest threat may not be the scientific illiteracy of the public, but the political illiteracy of scientists."[51] David Blockstein put it more bluntly: "Without politics, there will be no biodiversity."[52]

## Modern Wildlife Science: From Bricks to Buildings

The method of science, noted Carl Sagan, is far more important than the findings of science.[53] But to the outsider, the workings of science may look puzzling. On the one hand, science features creativity and self-correction; on the other hand, science appears unstructured and unpredictable. The full impact of today's discoveries can be difficult to gauge. Important findings may lurk unforeseen—even by experts, even when the breakthroughs are imminent.

Past predictions are instructive. In 1937, President Roosevelt assembled a panel of esteemed senior scientists and engineers to provide advice on the most important technical and industrial innovations in the United States for the coming twenty-five years. While the panel cited agricultural developments (correctly) and synthetic gasoline (wrongly), what they missed is telling. They overlooked antibiotics, jet aircraft, rockets, space exploration, computer advancement, the transistor, the laser, and genetic engineering.[54] In hindsight, these were no minor omissions.

This is the paradox of science. As Marc Kirschner pointed out, only retrospectively do we see the full implications of discovery: "One may be able to recognize good science as it happens, but significant science can only be viewed in the rearview mirror. . . . DNA restriction enzymes, once the province of obscure microbiological investigation, ultimately enabled the entire recombinant DNA revolution. Measurement of the ratios of heavy and light isotopes of oxygen, once a limited area of geochemistry, eventually allowed the interpretation of prior climate change."[55]

How, then, does one build an effective edifice?

History shows that we cannot know the future; we cannot easily sift through the unmade discoveries to pick out those most useful.[56] There is no straightforward, centralized, predestined recipe for significant science. Rather, the key is to assemble the ingredients—the right people, the right environment—bolstered by society's desire for diversified and productive research, however unruly and unpredictable it may seem.

Basic research creates unforeseen opportunities; it can help resolve impending issues. In the words of one astute Canadian researcher, "Emerging problems are more easily dealt with when basic research has already been done. In recent years, our ability to respond to problems, such as the mountain pine beetle or SARS, has depended on pre-existing research programs."[57] Wildlife science in North America provides its own examples. Here are just two: predicting the poleward shifts in species distributions in the face of climate change was derived from longstanding biogeographic studies of postglacial recolonization;[58] identifying critical habitat for threatened woodland caribou was assembled from scattered studies of habitat selection and demography.[59] In both cases, the groundwork—the whole significance of which may have been unforeseen—had begun years, even decades, in advance.

Wildlife science is anticipatory, poised to contribute in the Age of the Environment—a future where social concern about the environment will only grow. Consider biodiversity, invasive plants and animals, species niches and distributions: these are not only longstanding questions for wildlife ecologists; they are linked to issues of growing, collective concern including the transmission of infectious diseases, international trade and the breakdown of biogeographic barriers, and the effects of climate change and translocation of species.[60] Success in dealing with these matters will hinge on solid scientific understanding. These and other wildlife studies represent part of the blueprint for the next sturdy storey in the edifice of wildlife science. But the mortar of an informed public will be key.

## The Communication Imperative

Appreciating the importance of conveying conservation science is not new. The recognition is as old as conservation science itself.[61] Burnett wrote: "From the outset, it was evident to Harrison Lewis [head of the nascent Dominion Wildlife Service] that any hope for success in conserving Canada's wildlife would depend to a large degree on success in fostering public awareness and understanding."[62] From numerous lectures and media interviews to the lay language and the color photography of Canadian Wildlife Service *Reports*, to the broad appeal of *Hinterland Who's Who*, the public was regarded as a vital audience. It is an enduring imperative, and not just for the goal-directed endeavor of conservation biology. Recently, the Royal Society of Canada underscored that "the legitimacy of scientific evidence depends critically on communication and transparency."[63]

Back in the ivory tower, meanwhile, most researchers stay focused on communicating among themselves and educating the next generation of scientists to do much the same.[64] Detached from public discourse, yet conscious of the decline of nature, many ecologists seem resigned to be "alone in a world of wounds"—Leopold's smarting phrase.[65] Bridging this divide, some of my colleagues would argue, is merely a matter of time and patience. After all, our scientific findings are published, adrift somewhere on the sea of knowledge, someday to seep into the public consciousness. This slow mode of dissemination, however, fails in one crucial aspect: it's slow. Because of the accelerating pace of environmental change, "ecology is a discipline with a time limit."[66] Margaret Palmer and colleagues underscored the urgency: "We can no longer simply wait for our collective knowledge to be discovered by those who do not even know they need it."[67] There are plenty of excellent resources to assist scientists in climbing down from the ivory tower.[68]

## Conclusion: Into the Twenty-First Century

There is an old axiom in wildlife management, a lesson learned by every student of conservation: wildlife management is people management. Effective wildlife science entails more than biology. Indeed, Aldo Leopold was unwilling to treat the liberal arts and natural sciences as separate spheres, "not necessarily . . . because it was bad for wildlife, but because it was bad for people."[69] In the words of Nobel laureate John Polanyi, "If science is to affect the course of history, it must influence, not an equation or a computer, but people."[70] One species, *Homo sapiens*, will always be central to conservation.

More than a century after its emergence, the North American Model continues to serve wildlife by serving people. It is a vehicle for communicating, collaborating, and sharing experiences among professionals and sportsmen. One of the outcomes is trust. The involvement of hunters has been paramount—not just in studies of hunting, but in furnishing observations and samples crucial to generating knowledge at the heart of biological conservation: population size and trends, genetic connectedness, population limiting factors, and the spread of disease and invasive species.[71] In Newfoundland and Labrador, for instance, licensed caribou hunters have for decades submitted jawbones of the animals they harvested. This massive assemblage of samples, numbering in the thousands—each carefully measured for size, classified by age and sex—was key to understanding the decades-long rise and fall of Newfoundland caribou. These samples unveiled the changes in body size and the premature tooth wear consistent with food limitation of this island population, now a conservation unit of special concern.[72] Without hunters, without the long lens of time, this evidence would have remained unseen and unknown.[73]

Ecologists appreciate the importance of long-term data—the simple matter that "recognition of change and understanding the causes of change require long-term investment."[74] Indeed, change may dominate this century. The North American Model represents one avenue to gain that long-term perspective.

Hunters continue to support the science that supports conservation. As a government biologist, I learned that they are part of a creative liaison, at the front lines of conveying ideas and concerns with scientists. And in an urbanizing age, hunting may be instrumental in stemming the growing deficit of experiences by people in nature.[75] The North American Model is not flawless. It is, after all, an expression of the modern culture that engendered it and that was changed by it. Its scientific successes, nevertheless, underscore the importance of evidence, communication, collaboration, and trust—some very twenty-first century matters. For the foreseeable future, the hunting and the science at the crux of the North American Model are likely to be more relevant than ever.

*NOTES*

1. Royal Society of Canada, *Strengthening Government by Strengthening Scientific Advice: Fully Realizing the Value of Science to Canadian Society* (Ottawa: Royal Society of Canada, 2015).
2. C. Dean, *Am I Making Myself Clear? A Scientist's Guide to Talking to the Public* (Cambridge, MA: Harvard University Press, 2009).
3. P. B. Medawar, *The Limits of Science* (New York: Harper and Rowe, 1984).
4. J. G. Robinson, "Conservation Biology and Real-World Conservation," *Conservation Biology* 20, no. 3 (2006): 658–669.
5. V. Geist, "North American Policies of Wildlife Conservation," in *Wildlife Conservation Policy*, ed. V. Geist and Ian McTaggart-Cowan (Calgary: Detselig Enterprises, 1994), 77–129.
6. A. T. Knight, "Is Conservation Biology Ready to Fail?" *Conservation Biology* 23, no. 3 (2009): 517.
7. J. A. Schaefer and P. Beier, "Going Public: Scientific Advocacy and North American Wildlife Conservation," *International Journal of Environmental Studies* 70, no. 3 (2013): 429–437.
8. J. R. Platt, "Strong Inference," *Science* n.s. 146, no. 3642 (1964): 347–353.
9. P. R. Ehrlich. "Human Natures, Nature Conservation, and Environmental Ethics: Cultural Evolution Is Required, in Both the Scientific Community and the Public at Large, to Improve Significantly the Now Inadequate Response of Society to the Human Predicament," *BioScience* 52, no. 1 (2002): 31–43.
10. A. J. Hoffman, "Reflections: Academia's Emerging Crisis of Relevance and the Consequent Role of the Engaged Scholar," *Journal of Change Management* 16, no. 2 (2016): 77–96.
11. K. A. Artelle, J. D. Reynolds, A. Treves, J. C. Walsh, P. C. Paquet, and C. T. Darimont, "Hallmarks of Science Missing From North American Wildlife Management," *Science Advances* 4, no. 3 (2018): eaao0167; but see also J. R. Mawdsley, J. F. Organ, D. J. Decker, A. B. Forstchen, R. J. Regan, S. J. Riley, M. S. Boyce, J. E. McDonald, C. Dwyer, and S. P. Mahoney, "Artelle et al. (2018) Miss the Science Underlying North American Wildlife Management," *Science Advances* 4, no. 10 (2018): eaat8281.

12. W. J. Sutherland, A. S. Pullin, P. M. Dolman, and T. M. Knight, "The Need for Evidence-Based Conservation," *Trends in Ecology and Evolution* 19, no. 6 (2004): 305–308.
13. V. J. Meretsky and R. L. Fischman, "Learning from Conservation Planning for the U.S. National Wildlife Refuges," *Conservation Biology* 28, no. 5 (2014): 1415–1427.
14. C. N. Cook, M. Hockings, and R.W. Carter, "Conservation in the Dark? The Information Used to Support Management Decisions," *Frontiers in Ecology and Environment* 8, no. 4 (2009): 181–186.
15. R. L. Fischman and J. B. Rule, "Judging Adaptive Management Practices of U.S. Agencies," *Conservation Biology* 30, no. 2 (2016): 268–275.
16. National Science Board, "Science and Engineering Indicators 2012," National Science Foundation, January 2012, https://www.nsf.gov/statistics/seind12/.
17. T. Dobzhansky, "Nothing in Biology Makes Sense Except in the Light of Evolution," *American Biology Teacher* 35, no. 3 (1973): 125–129; J. D. Miller, E. C. Scott, and S. Okamoto, "Public Acceptance of Evolution," *Science* 313, no. 5788 (2006): 765–766; R. Dube, "What Does that Darwin Know Anyway?" *Globe and Mail*, June 19, 2007.
18. B. Spellberg, R. Guidos, J. Bradley, H. W. Boucher, W. M. Scheld, J. G. Bartlett, and J. Edwards Jr., "The Epidemic of Antibiotic-Resistant Infections: A Call to Action for the Medical Community from the Infectious Diseases Society of America," *Clinical Infectious Diseases* 46, no. 2 (2008): 155–164; J. Hemingway, N. J. Hawkes, L. McCarroll, and H. Ranson, "The Molecular Basis of Insecticide Resistance in Mosquitoes," *Insect Biochemistry and Molecular Biology* 34, no. 7 (2004): 653–655; D. W. Coltman, P. O'Donoghue, J. T. Jorgenson, J. T. Hogg, C. Strobeck, and M. Festa-Bianchet, "Undesirable Evolutionary Consequences of Trophy Hunting," *Nature* 426 (2003): 655–658; C. T. Darimont, S. M. Carlson, M. T. Kinnison, P. C. Paquet, T. E. Reimchen, and C. C. Wilmers, "Human predators Outpace Other Agents of Trait Change in the Wild," *Proceedings of the National Academy of Sciences of the United States of America* 106, no. 3 (2009): 952–954.
19. G. Ceballos, P. R. Ehrlich, A. D. Barnosky, A. Garcia, R. M. Pringle, and T. M. Palmer, "Accelerated Modern Human-Induced Species Losses: Entering the Sixth Mass Extinction," *Science Advances* 1, no. 5 (2015), doi: 10.1126/sciadv.1400253.
20. M. Hoffman, J. W. Duckworth, K. Holmes, D. P. Mallon, A. S. L. Rodrigues, and S. N. Stuart, "The Difference Conservation Makes to Extinction Risk of the World's Ungulates," *Conservation Biology* 29, no. 5 (2015): 1303–1313; M. A. Palmer, E. S. Bernhardt, E. A. Chornesky, S. L. Collins, A. P. Dobson, C. S. Duke, B. G. Gold, R. B. Jacobson, S. E. Kingsland, R. H. Kranz, M. J. Mappin, M. L. Martinez, F. Micheli, J. L. Morse, M. L. Pace, M. Pascual, S. S. Palumbi, O. Reichman, A. R. Townsend, and M. G. Turner, "Ecological Science and Sustainability for the 21st Century," *Frontiers in Ecology and Environment* 3, no. 1 (2005): 4–11; N. Baron, *Escape from the Ivory Tower* (Washington, DC: Island Press, 2010); Schaefer and Beier, "Going Public."
21. M. O'Regan, "Loss of Innocence as Third of Children Never Climbed a Tree," *Independent, Irish News*, March 6, 2016, https://www.independent.ie/irish-news/loss-of-innocence-as-third-of-children-never-climbed-a-tree-34515030.html.
22. K. van Mensvoort, "Children's Dictionary Dumps 'Nature' Words," *Next Nature*, April 2, 2009, https://www.nextnature.net/2009/02/childrens-dictionary-dumps-nature-words/.
23. A. R. Yott, R. Rosatte, J. A. Schaefer, J. Hamr, and J. Fryxell, "Movement and Spread of a Founding Population of Reintroduced Elk (*Cervus elaphus*) in Ontario, Canada," *Restoration Ecology* 19, no. 101 (2011): 70–77; S. H. M. Butchart, J. P. W. Scharlemann, M. I. Evans, S. Quader, S. Aricò, J. Arinaitwe, M. Balman, L. A. Bennun, B. Bertzky, C. Besançon, T. M. Boucher, T. M. Brooks, I. J. Burfield, N. D. Burgess, S. Chan, R. P. Clay, M. J. Crosby, N. C. Davidson, N. De Silva, C. Devenish, G. C. L. Dutson, D. F. Díaz Fernández, L. D. C. Fishpool, C. Fitzgerald, M. Foster, M. F. Heath, M. Hockings, M. Hoffmann, D. Knox, F. W. Larsen, J. F. Lamoreux, C. Loucks, I. May, J. Millett, D. Molloy, P. Morling, M. Parr, T. H. Ricketts, N. Seddon, B. Skolnik, S. N. Stuart, A. Upgren, and S. Woodley, "Protecting Important Sites for Biodiversity Contributes to Meeting Global Conservation Targets," *PLoS ONE* 7, no. 3 (2012), https://doi.org/10.1371/journal.pone.0032529; J. M. Reed, C. S. Elphick, E. N. Ieno, and A. F. Zuur, "Long-Term Population Trends of Endangered Hawaiian Waterbirds," *Population Ecology* 53 (2011): 473–481; Hoffman et al., "The Difference Conservation Makes."
24. R. P. Young, M. A. Hudson, A. M. R. Terry, C. G. Jones, R. E. Lewis, V. Tatayah, N. Zuël, and S. H. M. Butchart, "Accounting for Conservation: Using the IUCN Red List to Evaluate the Impact of a Conservation Organization," *Biological Conservation* 180 (2014): 84–96.
25. V. Geist, "An Introduction," in *Wildlife Conservation Policy*, ed. V. Geist and I. McTaggart-Cowan (Calgary: Detselig Enterprises, 1994), 7–23.
26. W. Davis. "Ecological Amnesia," in *Memory*, ed. P. Tortell, M. Turin, and M. Young (Vancouver: Peter Wall Institute for Advanced Studies, 2018), 21–29.

27. P. L. Errington and F. N. Hamerstrom, "The Evaluation of Nesting Losses and Juvenile Mortality of the Ring-Necked Pheasant," *Journal of Wildlife Management* 1, no. 1/2 (1937): 3–20.
28. J. Foster, *Working for Wildlife: The Beginning of Preservation in Canada*, 2nd ed. (Toronto: University of Toronto Press, 1998); J. A. Burnett, *A Passion for Wildlife: The History of the Canadian Wildlife Service* (Vancouver: University of British Columbia Press, 2002).
29. A. C. Isenberg, *The Destruction of the Bison: An Environmental History, 1750–1920* (Cambridge, UK: Cambridge University Press, 2000).
30. C. Meine, *Correction Lines* (Washington, DC: Island Press, 2004).
31. Foster, *Working for Wildlife.*
32. Geist, "North American Policies."
33. Foster, *Working for Wildlife.*
34. Foster, *Working for Wildlife.*
35. Foster, *Working for Wildlife.*
36. Foster, *Working for Wildlife.*
37. Foster, *Working for Wildlife.*
38. Burnett, *A Passion for Wildlife.*
39. Burnett, *A Passion for Wildlife.*
40. Geist, "North American Policies."
41. Burnett, *A Passion for Wildlife.*
42. C. Berger, "Book Review: Working for Wildlife," *Canadian Historical Review* 60 (1979): 85–86.
43. Burnett, *A Passion for Wildlife.*
44. Schaefer and Beier, "Going Public."
45. Burnett, *A Passion for Wildlife.*
46. Burnett, *A Passion for Wildlife.*
47. J. A. Hutchings and N. C. Stenseth, "Communication of Science Advice to Government," *Trends in Ecology & Evolution* 31, no. 1 (2016): 7–11.
48. A. Ø. Mooers, L. R. Prugh, M. Festa-Bianchet, and J. A. Hutchings, "Biases in Legal Listing under Canadian Endangered Species Legislation," *Conservation Biology* 21, no. 3 (2007): 572–575.
49. B. Owens, "Canada Names New Chief Science Adviser," *Science*, September 26, 2017, http://www.sciencemag.org/news/2017/09/canada-names-new-chief-science-adviser; Hutchings and Stenseth, "Communication of Science Advice."
50. Geist, "An Introduction."
51. G. E. van der Vink, "Scientifically Illiterate vs. Politically Clueless," *Science* 276, no. 5316 (1997): 1175.
52. D. E. Blockstein, "How to Lose Your Political Virginity while Keeping Your Scientific Credibility," *Bioscience* 52, no. 1 (2002): 91–96.
53. C. Sagan, *The Demon-Haunted World: Science as a Candle in the Dark* (New York: Random House, 1995).
54. C. H. Townes, "Unpredictability in Science and Technology," in *Science and Society*, ed. M. Moskovits (Concord, Ontario: House of Anansi Press, 1995), 29–42.
55. M. Kirschner, "A Perverted View of 'Impact,'" *Science* 340, no. 6138 (2013): 1265.
56. J. C. Polanyi, "Understanding Discovery," in *Science and Society*, ed. M. Moskovits (Concord, Ontario: House of Anansi Press, 1995), 3–12; Townes, "Unpredictability in Science."
57. Natural Sciences and Engineering Research Council of Canada, "Report of the International Review Committee on the Discovery Grants Program," January 14, 2010, http://www.nserc-crsng.gc.ca/_doc/Reports-Rapports/Consultations/international_review_eng.pdf.
58. M. B. Davis, "Lags in Vegetation Response to Greenhouse Warming," *Climatic Change* 15, no. 1–2 (1989): 75–82.
59. Environment Canada, *Scientific Assessment to Inform the Identification of Critical Habitat for Woodland Caribou (Rangifer tarandus caribou), Boreal Population, in Canada: 2011 Update* (Ottawa, Ontario: Environment Canada, 2011).
60. F. Keesing, L. K. Belden, P. Daszak, A. Dobson, C. D. Harvell, R. D. Holt, P. Hudson, A. Jolles, K. E. Jones, C. E. Mitchell, S. S. Myers, T. Bogich, and R. S. Ostfeld, "Impacts of Biodiversity on the Emergence and Transmission of Infectious Diseases," *Nature* 468 (2010): 647–652; D. Simberloff, J. Martin, P. Genoves, V. Maris, D. A. Wardle, J. Aronson, F. Courchamp, B. Galil, E. Garcia-Berthou, M. Pascal, P. Pyšek, R. Sousa, E. Tabacchi, and M. Vilà, "Impacts of Biological Invasions: What's What and the Way Forward," *Trends in Ecology and Evolution* 28, no. 1 (2013): 58–66; C. Parmesan and G. Yohe, "A Globally Coherent Fingerprint of Climate Change Impacts across Natural Systems," *Nature* 421 (2003): 37–42; D. Murray, Y. N. Majchrzak, M. J. L. Peers, M. Wehtje, C. Ferreira, R. S. A. Pickles, J. R. Row, and D. H. Thornton, "Potential Pitfalls of Private Initiatives in Conservation Planning: A Case Study from Canada's Boreal Forest," *Biological Conservation* 192 (2015): 174–180.
61. Schaefer and Beier, "Going Public."
62. Burnett, *A Passion for Wildlife.*
63. Royal Society of Canada, *Strengthening Government.*
64. Schaefer and Beier, "Going Public."
65. A. Leopold, *A Sand County Almanac: With Other Essays on Conservation from Round River* (New York: Oxford University Press, 1966).
66. F. Bazzaz, G. Ceballos, M. Davis, P. R. Ehrlich, T. Eisner, S. Levin, J. H. Lawton, J. Lubchenco, P. A.

Matson, H. A. Mooney, P. H. Raven, J. E. Roughgarden, J. Sarukhan, D. G. Tilman, P. Vitousek, B. Walker, D. H. Wall, E. O. Wilson, and G. M. Woodwell, "Ecological Science and the Human Predicament," *Science* 282, no. 5390 (1998): 879.

67. M. A. Palmer, E. S. Bernhardt, E. A. Chornesky, S. L. Collins, A. P. Dobson, C. S. Duke, B. D. Gold, R. B. Jacobson, S. E. Kingsland, R. H. Kranz, M. J. Mappin, M. L. Martinez, F. Micheli, J. L. Morse, M. L. Pace, M. Pascual, S. S. Palumbi, O. J. Reichman, A. R. Townsend, and M. G. Turner, "Ecological Science and Sustainability for the 21st Century," *Frontiers in Ecology and the Environment* 3 (2005): 4–11.
68. Dean, *Am I Making Myself Clear?*; Baron, *Escape from the Ivory Tower*; M. F. Dahlstrom, "Using Narratives and Storytelling to Communicate Science with Nonexpert Audiences," *Proceedings of the National Academy of Sciences* 111 (2014): 13614–13620.
69. Meine, *Correction Lines.*
70. Polanyi, "Understanding Discovery."
71. C. Godwin, J. A. Schaefer, B. R. Patterson, and B. A. Pond, "Contribution of Dogs to White-Tailed Deer Hunting Success," *Journal of Wildlife Management* 77, no. 2 (2013): 290–296; J. Novak, K. T. Scribner, W. D. Dupont, and M. H. Smith, "Catch-effort Estimation of White-Tailed Deer Population Size," *Journal of Wildlife Management* 55, no. 1 (1991): 31–38; H. G. Broders, S. P. Mahoney, W. A. Montevecchi, and W. S. Davidson, "Population Genetic Structure and the Effect of Founder Events on the Genetic Variability of Moose, *Alces alces*, in Canada," *Molecular Ecology* 8, no. 8 (1999): 1309–1315; A. M. Andreasen, "Identification of Source-Sink Dynamics in Mountain Lions of the Great Basin," *Molecular Ecology* 21, no. 23 (2012): 5689–5701; K. R. Lang and J. A. Blanchong, "Population Genetic Structure of White-Tailed Deer: Understanding Risk of Chronic Wasting Disease Spread," *Journal of Wildlife Management* 76, no. 4 (2012): 832–840; S. A. Seidel, C. E. Comer, W. C. Conway, R. W. Deyoung, J. B. Hardin, and G. E. Calkins, "Influence of Translocations on Eastern Wild Turkey Population Genetics in Texas," *Journal of Wildlife Management* 77, no. 6 (2013): 1221–1231.
72. S. P. Mahoney, J. N. Weir, J. G. Luther, J. A. Schaefer, and S. F. Morrison, "Morphological Change in Newfoundland Caribou: Effects of Abundance and Climate," *Rangifer* 31, no. 1 (2011): 21–34; J. A. Schaefer, S. P. Mahoney, J. N. Weir, J. G. Luther, and C. E. Soulliere, "Decades of Habitat Use Reveal Food Limitation of Newfoundland Caribou," *Journal of Mammalogy* 97, no. 2 (2016): 386–393; Committee on the Status of Endangered Wildlife in Canada, *Assessment and Status Report on the Caribou Rangifer tarandus, Newfoundland Population, Atlantic-Gaspésie Population, Boreal Population in Canada* (Ottawa: COSEWIC, 2014).
73. Despite its obvious value for generating knowledge and enhancing connections with the public, this program for hunter-submitted caribou and moose jawbones was terminated by the provincial government in 2016. It is a reminder that public spending on science will always be under scrutiny and will have to contend with other societal priorities.
74. D. B. Lindenmayer, E. L. Burns, P. Tennant, C. R. Dickman, P. T. Green, D. A. Keith, D. J. Metcalfe, J. Russell-Smith, G. M. Wardle, D. Williams, K. Bossard, C. deLacey, I. Hanigan, C. M. Bull, G. Gillespie, R. J. Hobbs, C. J. Krebs, G. E. Likens, J. Porter, and M. Vardon, "Contemplating the Future: Acting Now on Long-Term Monitoring to Answer 2050's Questions," *Austral Ecology* 40, no. 3 (2015): 213–224.
75. Lindenmayer et al., "Contemplating the Future."

Shane P. Mahoney

# North American Waterfowl Management

## *An Example of a Highly Effective International Treaty Arrangement for Wildlife Conservation*

*Consumptive use of migratory water birds in North America has a history likely dating back to the continent's earliest human cultures, and waterfowl remain significant to contemporary indigenous and nonindigenous peoples. This cultural tradition and our long history of subsistence and recreational use help explain the enduring value we attribute to ducks, geese, and swans relative to many other bird species. Furthermore, the original abundance, widespread declines, and subsequent recovery of this highly diverse assemblage of species represents one of the great stories in North American conservation. Perhaps more than any other wildlife resource, the management of the continent's waterfowl exemplifies the very best in our capacities to realize conservation success across massive geographic scales and sociopolitical complexity. Certainly, no other signature of the North American Model more explicitly defines how recreational hunting has served as the influencing force behind wildlife conservation and definitively represents the principle that wildlife must be scientifically managed as an international resource.*

### Introduction: The Early Steps for Waterfowl Protection

Influenced by a European economy hungry for exotics, a powerful domestic market, and a naive notion that the continent's wildlife resources were inexhaustible, early postcolonial North America was subject to a prolonged period of unregulated natural resource harvest that decimated many species including the iconic plains bison (*Bison bison bison*) and passenger pigeon (*Ectopistes migratorius*). By the mid-1800s, US naturalists, including J. J. Audubon and G. P. Marsh, began to vocalize their shared belief that wildlife, and migratory birds in particular, were fast declining and would require protection from excessive exploitation.[1] These concerns would increase in the early days of the twentieth century and be echoed by the Canadian conservation community.

This major conservation crisis ultimately resulted in the continent's first wave of coordinated conservation practice, a productive melange of utilitarian social values, wildlife harvest regulation, and wilderness protection that would eventually help formalize the concept of public trust resources, the heart of the North American Model.[2] This, in turn, helped guide the realization that because wildlife species paid no heed to political boundaries, the concept of the public trust itself must extend beyond political borders. Recognizing the special status of migratory wildlife, conservationists and decision-makers in North America began to consider and manage wildlife as an international resource. These concepts would eventually lead to precedent-setting international conservation agreements, one of the first being the 1911 North Pacific Fur Seal Convention, which was signed by the United States, Great Britain on behalf of Canada, Japan, and Russia and restricted international harvests out of concern for dwindling seal populations.

It was during this early conservation era that several important pieces of US legislation were established, including the Lacey Act of 1900 and the Weeks-MacLean Act of 1913, which together limited interstate commerce in dead wildlife, severely restricted market hunting, virtually outlawed spring hunting of birds altogether, and declared federal jurisdiction over migratory birds. Of special significance to the developing framework for sustainable use and management of North American wildlife, these legislations helped define legitimate purpose for wildlife harvest by eliminating commercial financial gain from dead wildlife as a legal practice and protecting species during the critical breeding season. Most relevantly, the Weeks-MacLean Act, known as the first Migratory Bird Act,[3] also set up an imperative to seek an international treaty on migratory birds to ensure the constitutional validity of US federal jurisdiction over migratory birds.[4] Unfortunately, a lack of funding and enforcement undermined this act, and in 1914 it was successfully challenged in a US district court (Arkansas) after questions arose regarding its constitutionality.[5]

Following the passage of the US Weeks-Maclean Act, Canadian conservation leaders sought to work jointly with their southern counterparts to establish an international agreement focused on the protection of migratory bird populations. They, too, recognized the challenges facing these species and the importance of international cooperation in sponsoring and developing effective legislation to regulate spring shooting, which disrupted the breeding cycle and was considered the main threat to North America's migratory birds at that time.[6] Eventually these cooperative efforts would bear fruit.

The resulting Migratory Bird Convention (Treaty) between the United States and Great Britain (on behalf of Canada) was signed in 1916, establishing an international agreement to protect useful and harmless migratory birds from indiscriminate killing and to adopt a shared system of cross-border protection. Drafting of the treaty actually began in 1914, but its completion was delayed by the outbreak of World War I. The chief negotiators were C. Gordon Hewitt of the Canadian Department of Agriculture and Edward Nelson of the US Biological Survey. Significantly, both men were scientists who approached the treaty negotiations through the lens of science and with a clear emphasis on conservation.[7] Mexico was experiencing significant political instability during this time and declined to join the first North American bird conservation discussions. Predictably, US states and Canadian provinces were initially reluctant to cede jurisdictional control of migratory birds to their national governments, but the treaty was eventually ratified, a testament to the emerging influence of public trust responsibilities within North America's nascent conservation framework and recognition of migratory birds as a transboundary public trust resource.

## An International Treaty for Migratory Birds

The Migratory Bird Convention (Treaty) necessitated and helped guide establishment of a federal bureaucracy responsible for migratory birds in Canada (eventually becoming the Canadian Wildlife Service [CWS]) and in the United States (eventually becoming the US Fish and Wildlife Service [USFWS]).[8] The treaty also established the basis for a uniform continental system of waterfowl harvest management, which is especially significant because most North American waterfowl are migratory. Though eventually requiring some revision to accommodate the seasonal and geographic nature of resource availability (including specifically the fact that Aboriginal and subsistence hunting in the north required access to waterfowl at different times of year than in southern regions), the Migratory Bird Convention (Treaty), in creating a uniform system for hunting seasons and harvest limits, was consistent with North America's emerging principles of resource management and the emphasis on sustainable use underpinnings in evaluating resource importance.[9] In particular, the treaty initially reflected and has since perpetuated the practice of ensuring democratic allocation of wildlife resources by law and preserving legal hunting opportunity for all North American citizens. From its inception, the Migratory Bird Convention (Treaty) acknowledged the original rationale for resource conservation in North America, which, though complex, was heavily steeped in utilitarian and human-benefit values.

Signing of the Migratory Bird Convention (Treaty) in 1916 was (despite early constitutional challenges in both countries) followed by passage of supportive domestic measures, including the Canadian Migratory Bird Convention Act (MBCA) of 1917 and the US Migratory Bird Treaty Act (MBTA) of 1918. The MBCA and MBTA prohibited take or possession of all listed migratory birds unless specifically allowed by hunting regulations, and hunting of designated game species was limited to specified open seasons between September 1 and March 10. These regulated harvests were to be based upon an informed review of population status and other pertinent criteria. With these measures adopted, the first broadly effective wildlife treaty in North America was officially in place.

Following the implementation of the Migratory Bird Convention (Treaty), conservation activists and policy-makers recognized the need to widen migratory bird protection efforts even further and responded with other international agreements.[10] For example, in 1936, the United States and Mexico negotiated a treaty to protect birds migrating between their two countries; in 1972, the United States and Japan signed a treaty to protect additional migratory bird species, enhance habitat, share scientific research data, and regulate hunting; and in 1976, a treaty was established with the Soviet Union to protect birds moving along flyways common to both regions.

The inclusion of wildlife in treaty law was a clear indication that the value of this resource was being recognized by legislators and bureaucrats at the highest government levels. It also demonstrated that North American conservation had matured past ensuring equal access to wildlife and the regulation of hunting privileges and practices. It had come to embrace the conservation of wild animals as a core ethic and a responsibility of national significance and a matter of national pride. This idea of transboundary responsibility and coordinated management of wildlife has become deeply ingrained in the North American Model of Wildlife Conservation and is responsible for a considerable array of ancillary policies, committees, colloquia, strategic approaches, and publications.[11]

The most enduring legal legacy of the 1916 treaty is the acknowledgment that migratory birds are held in public trust throughout North America and oversight responsibilities belong to the federal governments. Also significant is the degree to which the Migratory Bird Convention (Treaty) helped precipitate and focus the development of scientific underpinnings for wildlife management on the continent.[12] Perhaps no other single wildlife conservation effort so clearly defines the rationale and combined application of sustainable wildlife use and protection in North America. Furthermore, it is difficult to conceive how these early efforts in rescuing and protecting the continent's migratory bird resources could have emerged without recognition of their utilitarian value.

Despite the emerging conservation awareness that led to the Treaty, however, North America was at the same time experiencing a period of expansive industrialization and resource extraction that posed new and increasing threats to wildlife populations through disturbance and, most demonstrably, habitat loss. Waterfowl habitats, especially, were under strong development pressure with large-scale conversions of wetlands to agricultural use. Overexploitation, the haunting ghost of market hunting, was suddenly no longer the most pressing conservation concern; it was gradually being replaced by concern over habitat loss.[13]

## The Road to Waterfowl Science and Institution Building

These early, development-related challenges for waterfowl were exacerbated by the severe drought of the 1930s, which affected breeding grounds in Canada and the United States and resulted in drastic population declines. This additional crisis for waterfowl conservation, in turn, provoked a sudden array of creative conservation solutions that borrowed from and influenced the general wildlife conservation practices in place and being developed in North America.[14] At the time, natural history study was rapidly evolving toward the science of biology; and science, in general, was gaining greater social acceptance in North America and elsewhere. While many of the continent's conservation efforts continued to focus on wilderness protection and harvest regulation, a gradual shift toward science-based action, community and ecosystem-level approaches, and environmental regulation was beginning to occur.[15] Aldo Leopold and others significantly helped to advance this new paradigm, arguing in their writings and public lectures that an empirical approach to research and information-gathering was necessary to establish successful wildlife management policies.[16]

At the same time, there was considerable optimism about the capacity for humans to control nature for a greater good, including the restoration or improvement of nature. This resulted in enthusiastic efforts at predator reduction and citizen-attempted introductions and enhancement of desirable species in various areas of the continent.[17] Examples from waterfowl conservation included Jack Miner's Canada goose migration staging area in southern Ontario, created by persistently luring geese to his property until a population was essentially trained to make use of his refuge, and James Ford Bell's attempts to stock ducks on his property on the north shore of the Delta Marsh on Lake Manitoba because he wished to return more ducks to nature than he removed through his hunting activities.

However, not all efforts had the desired effect, and research was needed to explain why. It was quickly apparent to James Ford Bell, for example, that his Delta Marsh stocking efforts were failing to improve duck numbers, and in seeking expert advice, he convinced Aldo Leopold at the University of Wisconsin to send a graduate student to the Delta Marsh in 1937 to study waterfowl populations there.[18] With support from Bell, the Wildlife Management Institute and several private sponsors, the graduate student and (later) recognized waterfowl scientist, Albert Hochman, established the Delta Duck Station (known today as the Delta Waterfowl Foundation) to create a graduate student training program in partnership with dozens of North American universities.[19] Since then, the Delta Waterfowl Foundation has fostered international collaboration and trained generations of waterfowl biologists as managers, helping to further establish and advance professional expectations that waterfowl management and conservation be firmly grounded in science.

One of the earliest and most enduring developments in waterfowl research was the North American Bird Banding Program, started in the 1920s and continuing through today. Begun with a focus on determining species distributions and migration patterns, the effort increased in scope and soon necessitated a central system for standardizing and storing information. This service is provided today by the US Geological Survey's Bird Banding Laboratory, which maintains a central database. Early leg-banding efforts by Leon Cole, Jack Miner, and others affirmed and helped detail the cross-border movements of North American waterfowl.[20]

These first banding activities revealed that most waterfowl populations travel roughly north-south through their annual cycle, and this information provided a biological basis for coordinated management along political jurisdictions.[21] In 1935, Frederick Lincoln, a biologist with the Bureau of Biological

Survey, implemented the first midwinter inventory of waterfowl using aircraft and was also among the first to recognize the potential of using banding data for population estimation.[22] Through the late 1940s, the midwinter survey remained the primary source of information upon which hunting regulations and other policies requiring population status information were based.

Then, in 1955, the Waterfowl Breeding Population and Habitat Survey, a comprehensive aerial breeding ground survey for waterfowl across the north central United States, the southern prairie region of Canada, and vast sections of the eastern areas of both countries, became operational. In addition, by 1966, systematic North American Breeding Bird Surveys (which included waterfowl species) were being conducted collaboratively between the USFWS and CWS, adding further to our knowledge of migratory species abundance.[23] Since 1986, information gathered from this survey has been used to aid in monitoring progress toward population objectives derived under the North American Waterfowl Management Plan.[24]

One key and early achievement of the continent's waterfowl science effort was the delineation of the North American waterfowl flyways system. Although waterfowl leg-banding had been in use for decades, scientific analysis of band recovery data at a continental scale was not undertaken until Lincoln's work began in the 1930s.[25] Lincoln's identification and delineation of the major migration pathways of North American birds profoundly influenced waterfowl management policies for multiple jurisdictions. Prior to this effort, waterfowl management was applied exclusively within state and provincial borders; subsequently, Lincoln's flyways were used as the primary waterfowl management unit, adjusted, as necessary, for political boundaries to facilitate harvest regulations.[26]

Remarkably, these original flyway delineations have been repeatedly validated over time, despite enormous advances in waterfowl tracking technologies.[27] Eventually, Flyway Councils with membership from all states and provinces within a flyway would form the highly inclusive administrative system for waterfowl management that remains operational in North America today. These councils work with federal agencies to assist the development of annual frameworks for wider migratory game bird harvest regulations. In the United States, both flyway members and the general public provide input for the development of annual regulatory frameworks by the USFWS regulations committee. The regulatory frameworks are then forwarded to the secretary of the interior for approval. Parallel actions are taken by the CWS regulations committee. Canadian frameworks are subject to approval by the minister of environment and climate change.[28]

The importance of science as the basis for informed decision-making in wildlife conservation was recognized as critical and strongly promoted in these early waterfowl management efforts. The principle is ingrained today in the general North American approach to wildlife management.[29] The application of this principle has led to many advances in the management of diverse species, often under highly complex circumstances, including the adaptive management of migratory waterfowl harvests and waterfowl habitat conservation on a continental basis.[30] As these management systems matured, they, in turn, required and became motivating influences for more and improved science.

## Developing and Funding Habitat Conservation Programs for Waterfowl

Three developments in the 1930s proved especially valuable for waterfowl habitat conservation: creation of duck stamps, emergence of nongovernment organizations (NGOs) devoted to waterfowl research and wetland conservation, and growth of habitat-related research.[31]

The first migratory bird sanctuary in North America was established by Canada in 1887 at Last Mountain Lake, Saskatchewan, while the first US refuge, Pelican Island National Wildlife Refuge, was established in 1903.[32] Following the implementation of the Migratory Bird Convention (Treaty), funding was needed to support the associated legislative obligations, particularly for additional and expansive habitat projects. In 1934, the Federal Migratory Bird Hunting and Conservation Stamp (the Duck Stamp) was introduced in the United States. A stamp was produced and sold to all hunters, serving as their federal permit to hunt migratory waterfowl, and the revenue generated was originally earmarked for waterfowl refuges. The first Duck Stamp sold for $1.00. Between 1934 and 2012, US$866 million were generated by the US Duck Stamp program.[33] These funds were used to purchase or lease more than 2.4 million hectares of wetland habitat in the United States, including the protection of seven thousand Waterfowl Production Areas in the Prairie Pothole Region.[34]

Then, in 1937, the more general Pittman-Robertson Act became US law. It introduced a 10–11 percent manufacturer's excise tax on sporting firearms and ammunition, the revenues from which were to be dispersed to state wildlife agencies for the purpose of habitat restoration on a cost-share basis. By the mid-1980s, the Pittman-Robertson excise tax had generated US$1.5 billion and leveraged an addition US$500 million from the states and, by 2013, more than US$8 billion had been raised.[35] While Pittman-Robertson funds are available for and applied to a broad array of wildlife conservation efforts, they certainly have supported waterfowl conservation specifically and in a significant way.

While both these successful US funding strategies ensure public funding is available in support of waterfowl conservation efforts, they are specifically set up as a user-pay system, with users generally being defined as hunters. In the context of the North American Model, this arrangement exemplifies the supportive connections between various Model components, including ensuring conservation through sustainable resource use, the role of hunters in financially supporting conservation, and utilizing science as the basis for decision-making. What is often overlooked in discussions of sustainable wildlife use in North America, is how significant the hunting community has been in actually advancing wildlife science through the funds it generates for research, the biological information it provides, and the demands it places on management agencies to ensure viable wildlife populations and sustainable harvests.

Early Canadian efforts at funding waterfowl habitat programs were somewhat less successful. The mandatory Canada Migratory Game Bird Hunting Permit, first introduced in 1966, did not originally direct revenues to habitat conservation.[36] Additionally, programs initiated in the 1960s to secure habitat through conservation easements did not realize their goals, and, as a result, outright purchase of land soon became the new strategy for waterfowl habitat protection in Canada. This achieved some success, securing 18,000 hectares in thirty-four national wildlife areas, but ultimately did not meet habitat conservation objectives before the program was eliminated during federal budget cutbacks in 1984.[37]

In 1984, Canada's national Migratory Bird Hunting Permit, a program similar to the US federal Duck Stamp, was initiated to fund Wildlife Habitat Canada, a national, charitable, nonprofit conservation organization. Today, the Canadian permit must be affixed with a Canadian Wildlife Habitat Conservation Stamp, and, to date, more than 1,500 conservation projects across Canada have been supported through the sale of the stamp.[38] While the stamp is purchased primarily by waterfowl hunters to validate their Migratory Bird Hunting Permits, funded projects may focus on conservation of any species. However, Wildlife Habitat Canada is mandated to recognize the primary source of funding and therefore primarily supports proposals that address priority activities for waterfowl and migratory bird conservation, including human dimensions research; impacts on regional and local habitats of importance to other migratory game birds; and prioritizing activities under the Newfoundland and Labrador Murre Conservation Fund.[39] Some revenue is also directed toward hunter education and recruitment projects across Canada, since, Canada's waterfowl conservation strategy, like that of the United States, depends heavily on the success of the user-pay system.[40]

Indeed, the contribution of North American hunters historically to waterfowl conservation extends well beyond making payments for licenses, permits, and stamps. At the height of the early conservation movement in North America, and around the time the treaty was signed, highly organized NGOs, adept at fundraising from both private and public sources, were rapidly being established and were beginning to successfully protect and restore wild lands.[41] Many of these citizen groups were founded by hunters, sportsmen committed to funding, bolstering, and actively pursuing conservation efforts to perpetuate hunting through the protection and restoration of wildlife habitat. Waterfowlers, too, enthusiastically joined these ranks. One example was their founding in 1930 of the More Game Birds in America Foundation. Established by a group of American businessmen, this organization sponsored reconnaissance trips and surveys focused on the waterfowl breeding grounds in Canada. The purpose of these excursions was to determine the habitat impacts of the severe drought being experienced at the time and how these habitat damages could be reversed.[42] Based partly on findings from these trips, the urgent need for a unique conservation organization in support of waterfowl conservation was recognized and Ducks Unlimited (DU) was incorporated in the United States in 1937, followed by Ducks Unlimited Canada (DUC) in 1938, and Ducks Unlimited Mexico (DUM) in 1974.

DU and DUC have played a pivotal and high-profile role in collaborative efforts between the United States and Canada to conserve waterfowl. To achieve this collaboration, it was necessary for the US waterfowl conservation movement to acknowledge that since US hunters directly benefited from the expansive waterfowl breeding grounds in Canada, US hunters had a responsibility to support habitat protection, restoration, and management efforts in Canada. For example, early Canadian conservationists who perceived spring hunting as the main threat to waterfowl populations considered that it could be curtailed, in part, through domestic legislative efforts. However, until US hunters and policymakers also demonstrated a willingness to protect migratory birds in spring and the species' breeding habitat in Canada, Canada's solitary efforts were perceived as ineffectual and unproductive.[43]

Regardless of mutual recognition of problems facing waterfowl and the desire for affirmative action, however, no mechanism existed whereby US public funds could be directed to Canada for waterfowl conservation. As an NGO, DU provided a valuable service in this context, raising and leveraging funds (including US state wildlife agency funds) through its own mechanisms for habitat conservation in Canada.[44] Since 1937, DU, founded and led largely by hunters, has raised US$4.5 billion for conservation in the United States alone. By January 2017, it had conserved 6.4 million acres (2.6 million hectares) of waterfowl breeding and staging habitat in Canada, 5.5 million acres (2.2 million hectares) in the United States, and 1.9 million acres (769,000 hectares) in Mexico.[45]

Notably, other hunter-based NGOs and individual landowners have also made substantive contributions to waterfowl habitat conservation.[46] However, the roles DU and DUC have played in North American waterfowl conservation and in raising public awareness for this effort are outstanding by any measure.

Thus, even in advance of the North American Waterfowl Management Plan (discussed below), migratory bird management norms included a reliance on the concept of public trust, international cooperation, public-private partnership funding arrangements for habitat conservation, scientific data for decision-making, uniform harvest regulation at a near-continental scale, and regional input and implementation of management options. When the drastic waterfowl declines of the mid-1970s and consequent reduced opportunities for hunting occurred, the North American waterfowl program was therefore prepared to quickly respond, refining its strategies and refocusing its efforts. The long-established biologically and politically delineated regionalization of waterfowl management was thus a clear precursor, at least conceptually, to the next major innovation in waterfowl management, the habitat joint venture model known as the North American Waterfowl Management Plan (NAWMP).

## The North American Waterfowl Management Plan

The NAWMP was developed to address severe declines in waterfowl populations occurring across North America in the 1970s and was jointly drafted and endorsed by the US and Canadian governments in 1986, with Mexico becoming a signatory in 1994. The NAWMP is a framework for the continental-scale management of waterfowl population abundance primarily through aggressive habitat conservation and has been the catalyst for much of the wetland habitat conservation work in North America for more than thirty years.[47] Based on assessments of biological needs, human desires for hunting and other recreational use, and a principle of shared responsibility for the stewardship of waterfowl and their habitats, the NAWMP is a living document which was revised in 1994, 1998, 2004, and 2012.

Its basic principles and objectives have remained consistent, however. These include: ensuring diverse and abundant waterfowl populations are valued; recognizing that international cooperation and coordination are key to achieving the NAWMP vision; emphasizing that the scope of planning must be continental but implementation must be regional; realizing that habitat restoration and conservation are imperative to support waterfowl populations; promoting managed waterfowl harvest as consistent with conservation; and demonstrating that the pooling of government, conservation organization, and private interest resources is the most efficient model for funding the plan. These principles are fully in line with and supportive of previous migratory bird management approaches. So what innovation did this new program bring to North American waterfowl management?

When the NAWMP was first introduced in 1986, the scope of the plan, defined as management of thirty-seven waterfowl species across national, state, and provincial borders required regional implementation structures that did not yet exist. Furthermore, it required unprecedented, direct funding commitments to which the federal signatories could not pledge themselves. In essence, the plan called for a unique and daring new conservation effort led by (then) nonexistent regional joint ventures (JVs). These would provide the regional habitat stewardship needed to fulfil the annual-cycle requirements of waterfowl via voluntary partnerships between government and private interests.[48] Doing so would greatly expand the level of direct citizen engagement with waterfowl management in North America, but would require massive coordination and commitment at numerous levels across a varied assemblage of organizations. However, despite the scale of the challenge, the successful implementation of this plan has been remarkable, an example of North American conservation at its best.

### Program Delivery

The NAWMP program delivery relies primarily on the regional habitat JVs to execute actions supporting NAWMP goals for waterfowl abundance and habitat conservation within biologically delineated geographic regions in Canada, Mexico, and the United States. In addition, there are species JVs, all international in scope, which are designed to acquire the science needed to inform management decisions for priority species. The JVs are so called because they require collaborative participation by private and government organizations for funding research and management projects; in some cases, the collaborative participation is also necessary to put boots on the ground operationally.

The implementation mechanisms of the NAWMP form a complex web of distinct but interrelated parts, which are coordinated by federal agencies whose decision-making processes reflect the political and social values of three different nations. The strategic framework for the NAWMP is set by the NAWMP Plan Committee. Its membership is made up of delegates from all three signatory nations who represent federal governments, state/provincial governments, and key NGOs, particularly DU and DUC. Several working groups comprised of technical experts focus on issues such as communications, policy, and science, and support the Plan Committee in updating the workplan for the NAWMP every five years. Additional working groups support integration and coordination of NAWMP activities in the broader waterfowl management community (e.g., Interim Integration Committee, Human Dimensions Working Group, Public Engagement Team, and the Harvest Management Working Group).[49]

It was recognized early in the NAWMP effort that research and evaluation were vital not only in themselves but also for guiding efficient expenditures of the new conservation funds arising from the program.[50] A NAWMP Science Support Team (NSST), with membership from all signatory nations, facilitates scientific communication and collaborative research among the joint ventures, the Plan Committee, and the respective federal wildlife agencies. A primary focus of the NSST is to provide capacity and advice for adaptive management approaches to waterfowl conservation.[51] Although formal constructs of adaptive resource management had not yet appeared in waterfowl management when the NAWMP was formalized, stakeholders increasingly shared the idea that performance evaluation and adjustment were critical to the program's success. While such evaluations were slow to develop in many joint ventures due to limited funding in the early days of the approach, progress was gradually made on this and on all aspects of the NAWMP.[52]

## Achievements

By 1994, Mexico was a full partner in the NAWMP, joining with Canada and the United States. In addition, fourteen regional structures (twelve habitat JVs and two species joint ventures focused on research, one for black duck [*Anas rubripes*] and one for arctic-breeding geese species, such as Canada goose (*Branta* spp.) and snow goose (*Anser* spp.), were established; US$500 million had been contributed for wetlands and waterfowl conservation; and 2 million acres (810,000 hectares) of waterfowl habitat benefited from on-the-ground conservation efforts.[53] Waterfowl population abundance increased consistently with NAWMP objectives in the 1990s, though it is not entirely clear to what extent this resulted from these wetland conservation efforts.[54] Regardless, progress continued and by 2015, in Canada alone, a staggering 173.2 million acres (70 million hectares) of waterfowl habitat had been subjected to some degree of conservation effort and twenty-five (twenty-two habitat and three species) JVs were operating in North America.[55] These efforts united agencies, NGOs, and private citizens in a common cause, a cause first mobilized almost a century earlier and primarily funded throughout by the sustainable use community of North America.

## Funding Mechanisms

One of the unique characteristics of the NAWMP is that funding for the JVs flows across national borders. Recognizing that breeding and wintering habitat are vital to support and maintain abundant waterfowl, the United States invests substantially in conservation efforts in Canada and Mexico, respectively. Since 1986, more than C$1 billion from US sources has been contributed to NAWMP efforts in Canada and about US$20 million has been used to support 102 projects in thirty-one Mexican states.[56]

Thousands of individual voluntary public-private partnerships have contributed to making these projects and the NAWMP successful. The challenging strategy of seeking matching funds for common priorities across three countries, while difficult, eventually proved effective, especially with the introduction of the North American Wetlands Conservation Act (NAWCA) in 1989.[57] In that same year, the US North American Wetlands Conservation Act (NAWCA [US]) created a mechanism to use federal US funds to leverage additional contributions from public agencies (e.g., Canadian federal government, state governments), DU, other conservation NGOs, and corporations and to provide grants for the conservation of wetlands and associated habitat.

Leveraging is achieved by requiring grant proposals to demonstrate matching funds from sources other than US federal funds, and for Canadian projects to demonstrate that 25 percent of the funds are from Canadian sources.[58] The NAWCA also established the North American Wetlands Conservation Council (NAWCC), the body that makes funding recommendations based on proposals reviewed and ranked by JV management boards. The Migratory Bird Conservation Commission, with recommendations from NAWCC, then allocates funds in accordance with NAWCA guidelines, which decree that at least 30 percent and not more than 60 percent of the available funds are to be dispersed to Canada and Mexico.[59]

Through the local review and ranking of proposals by JV management boards, the NAWCA funding model allows for national priorities and cultural practices to be reflected in the funding allocations to each signatory country. The US grants are awarded based on a competitive process that considers JV rankings as well as other criteria, including the quality and quantity of matching funds from partner organizations. The Canada grants are awarded in a programmatic way, allocating funds by conservation priority (every year, the majority of the funds go to the Prairie Habitat Joint Venture), and proposals consistent with the NAWCC requirements and NAWMP principles are developed by agencies and partners to access the amount of money available.[60] Clearly the funding model is complex and multitiered and thus has many levels of oversight. These layers provide multiple opportunities for scrutiny to ensure that funding is used to support projects and activities that are focused on achieving NAWMP goals.

Not surprisingly, the NAWMP has not been entirely successful in meeting all its stated goals, particularly as these relate to population abundance for all waterfowl species.[61] However, the NAWMP has certainly succeeded in making significant progress toward many of its objectives and, consistent with the principle of adaptive management, the NAWMP continues to be a work in progress and subject to ongoing revision.[62] Perhaps most notably, the NAWMP has succeeded as a model for cooperative long-term international efforts toward common conservation objectives. In this regard, the NAWMP's complex, multi-stakeholder delivery network may be the key to its success.[63]

When first introduced, the NAWMP appeared as a new approach to conservation; since then, the program has been used as a model for many subsequent international conservation efforts.[64] The very characteristics that made the NAWMP appear new, however—the continental scale of its vision and the complex web of interrelated but distinct parts—were, in reality, an elaboration of a long history of international cooperation on migratory bird science and management and a formalized connectivity of many existing models, mechanisms, and programs already emergent within the North American Model of wildlife management and conservation.

Nevertheless, the NAWMP has been rightfully described as the most ambitious conservation initiative ever attempted in North America.[65] Considering the coordination required across multiple levels of national, state, and provincial governments to implement actions at several geographic scales, through the synergistic efforts of thousands of public-private partnerships and agreements, all designed to address the management of nearly forty species of duck, goose, and swan, that assessment may, in fact, be an understatement. The NAWMP is an exceptional conservation effort by any measure and remains an example of what can be achieved through long-term commitments to science-based and citizen-supported conservation.

It is also notable that some of the most significant early commitments to the NAWMP program evaluation were made by groups such as DU and the Delta Waterfowl Foundation. In fact, waterfowl NGOs, supported primarily by hunters, remain one of the largest funding sources for JV program overview and evaluation, not just for program development. This has been especially true for projects in Canada, where a strong effort has been made to ensure programs for enhancing waterfowl production and supported by sizeable expenditures of NAWMP funds are carefully assessed.[66]

## Reflections on How Success Was Achieved

As with most major policy innovations, changes in the way conservation is approached is usually precipitated by crisis.[67] The NAWMP was proximately motivated by a continent-wide decline in duck populations from about 100 million in the 1970s to 55 million in the mid-1980s. In the face of growing human population pressure and development, especially for agricultural land use, the idea that a continental effort navigating divergent international policies, priorities, and capacities—an effort requiring one country to contribute in large proportion to work required outside its own borders, all for the primary purpose of conservation of wildlife habitat—still feels novel. The mobilization of resources to meet ambitious conservation goals that were articulated even before the necessary mechanisms to implement them were in place would probably seem implausibly rapid to an outside observer. The speed of the success in implementation of the NAWMP is due in a large part to the groundwork laid by a long but steady history of international cooperation on migratory bird management, which developed concurrently with the establishment of a general framework for sustainable use of wildlife in North America.

Notably, Americans and Canadians were working collaboratively on migratory bird monitoring and scientific research at least since the late 1880s through such activities as participation in natural history organizations and professional associations, the Christmas Bird Count, and research activities such as the US government's Biological Survey field work in Canada.[68] Much of this activity was propelled by waterfowl hunters, who recognized the significance of the Canadian breeding grounds to the maintenance of diverse and abundant waterfowl populations. Indeed, international cooperation on migratory bird issues and on waterfowl in particular, was about a century old by the time the NAWMP was introduced.

The mechanisms for delivery of the NAWMP, or precursors of these mechanisms, were largely in place when negotiations between the United States and Canada began to establish an international agreement for joint resource management purposes.[69] Mexico was invited to negotiate the original NAWMP but declined until a better understanding of Mexico's role was developed. Initially, there was a lot of pressure for the lead negotiators (Harvey Nelson for the United States and James Patterson for Canada) to establish a formal agreement for harvest allocation before negotiating management planning, which was unattainable; but as soon as the focus shifted from waterfowl harvest to the conservation of waterfowl and wetlands, negotiations progressed quickly.[70] Instead of who had access to how many birds, the focus turned to action to support more birds.

Building on a long history of international cooperation and leaning on the North American tenets of sustainable use management, the NAWMP introduced some novel practices that evolved from successful practices of the past, such as creating a mechanism for US federal funds to be directed across borders, which borrowed the leveraging model already in use by DU. Joint ventures were able to localize conservation efforts within the existing regional flyways. The NAWMP also formalized the public-private funding model for habitat conservation activity and significantly helped to validate the importance of hunter-led efforts in wildlife conservation.

## Conclusion: Future Directions

Leading up to the 2012 NAWMP Update, waterfowl biologists and managers from all three countries participated in a broad review of the program, reexamining its original goals and intentions in the context of current knowledge and changing conservation values at a broad social scale.[71] While one of the fundamental goals of the 1986 NAWMP had been to recover waterfowl populations to satisfy the demand for hunting opportunity, when waterfowl populations did recover, hunting uptake did not follow as expected, implying that the valuation of waterfowl for people had indeed changed over time.[72] Gaps in connectivity between habitat restoration and population abundance and inadequacies in the evaluation of the NAWMP progress and success were also identified as weakness in the 2102 review.[73] Additionally, reviewers noted that North America appears to be entering a new conservation era characterized by landscape-level solutions, such as sustainable development, which has obvious implications for NAWMP approaches and deliverables; and that increased social desire for accountability of public spending has increased the use of strategic planning and evaluation within government agencies, processes that can slow delivery of complex programs like the NAWMP.

The 2012 NAWMP Update addressed these issues by shifting the approach to waterfowl management in significant ways. Concepts of ecosystem resilience, ecological services, and human dimensions were introduced to the modified NAWMP, consistent with landscape-level conservation. Similarly, principles of sustainable development were applied and the plan acknowledged that social and economic research should complement existing biological and ecological knowledge for decision-making. A new focus on the integration of population, habitat, and human dimension components, including both consumptive and nonconsumptive valuations for waterfowl, were also explicitly included. Furthermore, efforts to increase the relevancy of waterfowl to humans, or to understand a changed relevancy in this regard, were identified as necessary to ensure the continued engagement and support required for success in the public-private collaboration.[74]

This new focus is refreshing and indicates a willingness on the part of the continental waterfowl community to embrace innovation to improve continental conservation success. The renewed plan, however, also continues to build on the accomplishments of the past by adhering to principles that have resulted in its major successes: complexity, diversity, international collaboration and management, sound science, and strong support from North America's sustainable use community. These principles explicitly align with the wider North American Model of Wildlife Conservation.

*NOTES*

1. M. G. Anderson, D. Caswell, J. M. Eadie, J. T. Herbert, M. Huang, D. D. Humburg, F. A. Johnson, M. D. Koneff, S. E. Mott, T. D. Nudds, E. T. Reed, J. K. Ringelman, M. C. Runge, B. C. Wilson, "Report from the Joint Task Group for Clarifying North American Waterfowl Management Plan Population Objectives and Their Use in Harvest Management," Atlantic Coast Joint Venture, 2007, http://acjv.org/documents/Final_Report_of_the_Joint_Task_Group.pdf.
2. S. P. Mahoney, ed., *Conservation and Hunting in North America*, monograph issue, *International Journal of Environmental Studies* 72, no. 5 (2015): 731–899; J. F. Organ, V. Geist, S. P. Mahoney, S. Williams, P. R. Krausman, G. R. Batcheller, T. A. Decker, R. Carmichael, P. Nanjappa, R. Regan, R. A. Medellin, R. Cantu, R. E. McCabe, S. Craven, G. M. Vecellio, and D. J. Decker, "The North American Model of Wildlife Conservation," *Wildlife Society Technical Review* 12, no. 4 (2012): 25.
3. M. G. Anderson, R. T. Alisauskas, B. D. J. Batt, R. J. Blohm, K. F. Higgins, M. C. Perry, J. K. Ringelman, J. S. Sedinger, J. R. Serie, D. E. Sharp, D. L. Trauger, and C. K. Williams, "The Migratory Bird Treaty and a Century of Waterfowl Conservation," *Journal of Wildlife Management* 82, no. 2 (2018): 247–259.

4. J. Wilson, "'New' and 'Old' Modes of Environmental Governance: The Evolution of the North American Waterfowl Bird Policy Regime," Western Political Science Association Conference, Albuquerque, NM, March 16–18, 2006.
5. Anderson et al., "Migratory Bird Treaty."
6. Anderson et al., "Migratory Bird Treaty."
7. Wilson, "'New' and 'Old" Modes of Environmental Governance."
8. F. Cooke, "Ornithology and Bird Conservation in North America—A Canadian Perspective," *Bird Study* 50, no. 3 (2003): 211–222.
9. M. G. Anderson and P. I. Padding, "The North American Approach to Waterfowl Management: Synergy of Hunting and Habitat Conservation," *International Journal of Environmental Studies* 72, no. 5 (2015): 810–829; V. Geist, "North American Policies of Wildlife Conservation," in *Wildlife Conservation Policy*, ed. V. Geist and I. McTaggart-Cowan (Calgary: Detselig Enterprises, 1995), 75–129; V. Geist, S. P. Mahoney, and J. F. Organ, "Why Hunting Has Defined the North American Model of Wildlife Conservation," *Transactions of the North American Wildlife and Natural Resources Conference* 66 (2001): 175–185.
10. Anderson et al., "Migratory Bird Treaty," 247–259.
11. P. J. Dart, R. Keck, M. C. Bambery, G. R. Batcheller, G. DeGayner, D. Fielder, V. Geist, D. Hobbs, J. E. Kennamer, S. P. Mahoney, J. E. McDonald Jr., J. F. Organ, R. Regan, and R. D. Sparrowe, "North American Model of Wildlife Conservation," Sporting Conservation Council, September 7, 2008, https://www.peer.org/assets/docs/doi/08_9_7_sporting_council_white_papers.pdf.
12. Anderson et al., "Migratory Bird Treaty."
13. Anderson and Padding, "North American Approach."
14. Geist, "North American Policies."
15. Mahoney, *Conservation and Hunting*.
16. Anderson et al., "Migratory Bird Treaty."
17. Mahoney, *Conservation and Hunting*.
18. Anderson and Padding, "North American Approach."
19. Anderson and Padding, "North American Approach."
20. US Geological Survey, "Bird Banding Laboratory," April 27, 2016, https://www.usgs.gov/centers/pwrc/science/bird-banding-laboratory?qt-science_center_objects=0#qt-science_center_objects.
21. Anderson et al., "Migratory Bird Treaty."
22. A. S. Hawkins, *Flyways: Pioneering Waterfowl Management in North America* (Washington, DC: United States Department of the Interior, Fish and Wildlife Service, 1984); Anderson et al., "Migratory Bird Treaty."
23. Cooke, "Ornithology and Bird Conservation."
24. J. R. Sauer, J. E. Fallon, and R. Johnson, "Use of North American Breeding Bird Survey Data to Estimate Population Change for Bird Conservation Regions," *Journal of Wildlife Management* 67, no. 2 (2003): 372–389.
25. Anderson and Padding, "North American Approach."
26. Wilson, "'New' and 'Old' Modes of Environmental Governance."
27. Anderson and Padding, "North American Approach."
28. Anderson et al., "Migratory Bird Treaty."
29. Geist, "North American Policies."
30. J. D. Nichols, F. A. Johnson, and B. K. Williams, "Managing North American Waterfowl in the Face of Uncertainty," *Annual Review of Ecology and Systematics* 26 (1995): 177–199.
31. Anderson et al., "Migratory Bird Treaty."
32. Anderson et al., "Migratory Bird Treaty."
33. Anderson and Padding, "North American Approach."
34. Anderson et al., "Migratory Bird Treaty."
35. Wilson, "'New' and 'Old' Modes of Environmental Governance"; Anderson and Padding, "North American Approach."
36. Wilson, "'New' and 'Old' Modes of Environmental Governance."
37. Wilson, "'New' and 'Old' Modes of Environmental Governance."
38. Government of Canada, Environment and Natural Resources, "Canadian Wildlife Habitat Conservation Stamp," April 17, 2018, https://www.canada.ca/en/environment-climate-change/services/migratory-bird-conservation/canadian-wildlife-habitat-stamp.html.
39. Wildlife Habitat Canada, "What We Do," 2018, https://whc.org/what-we-do/.
40. Wildlife Habitat Canada, "What We Do."
41. Mahoney, *Conservation and Hunting*.
42. Anderson and Padding, "North American Approach."
43. Anderson et al., "Migratory Bird Treaty."
44. Wilson, "'New' and 'Old' Modes of Environmental Governance."
45. Ducks Unlimited, "Fact Sheet," February 5, 2018, https://www.ducks.org/media/_global/_documents/stateFactSheets/NationalFactSheet.pdf.
46. Anderson and Padding, "North American Approach."
47. Anderson et al., "Migratory Bird Treaty."
48. Anderson and Padding, "North American Approach."
49. US Fish and Wildlife Service, "North American Waterfowl Management Plan," October 4, 2016, https://www.fws.gov/birds/management/bird-management-plans/north-american-waterfowl-management-plan.php.
50. Anderson and Padding, "North American Approach."
51. US Fish and Wildlife Service, "North American Waterfowl."

52. Anderson and Padding, "North American Approach."
53. North American Waterfowl Management Plan Committee, *North American Waterfowl Management Plan, 1998 Update: Expanding the Vision* (Arlington, VA: North American Waterfowl and Wetlands Office, US Fish and Wildlife Service, 1999).
54. North American Waterfowl Management Plan Committee, *North American Waterfowl Management Plan 2012: People Conserving Waterfowl and Wetlands* (Ottawa: Canadian Wildlife Service, 2012); Anderson et al., "Migratory Bird Treaty."
55. Anderson and Padding, "North American Approach."
56. North American Waterfowl Management Plan Committee, *North American Waterfowl Management Plan 2012*.
57. Anderson and Padding, "North American Approach."
58. Wilson, "'New' and 'Old' Modes of Environmental Governance."
59. Wilson, "'New' and 'Old' Modes of Environmental Governance."
60. Wilson, "'New' and 'Old' Modes of Environmental Governance."
61. North American Waterfowl Management Plan Committee, *North American Waterfowl Management Plan 2012*; Wilson, "'New' and 'Old' Modes of Environmental Governance"; B. K. Williams, M. D. Koneff, and D. A. Smith, "Evaluation of Waterfowl Conservation under the North American Waterfowl Management Plan," *Journal of Wildlife Management* 63, no. 2 (1999): 417–440.
62. Anderson et al., "Report from the Joint Task Group."
63. Wilson, "'New' and 'Old' Modes of Environmental Governance."
64. US Fish and Wildlife Service, "North American Waterfowl;" Cooke, "Ornithology and Bird Conservation."
65. North American Waterfowl Management Plan Committee, *North American Waterfowl Management Plan, 1998 Update*.
66. Anderson and Padding, "North American Approach."
67. Mahoney, *Conservation and Hunting*; Wilson, "'New' and 'Old' Modes of Environmental Governance"; Anderson and Padding, "North American Approach."
68. Wilson, "'New' and 'Old' Modes of Environmental Governance."
69. Wilson, "'New' and 'Old' Modes of Environmental Governance."
70. Wilson, "'New' and 'Old' Modes of Environmental Governance."
71. North American Waterfowl Management Plan Committee, *North American Waterfowl Management Plan 2012*; Anderson et al., "Report from the Joint Task Group."
72. North American Waterfowl Management Plan Committee, *North American Waterfowl Management Plan 2012*.
73. Williams, Koneff, and Smith, "Waterfowl Conservation"; North American Waterfowl Management Plan Committee, *North American Waterfowl Management Plan 2012*.
74. North American Waterfowl Management Plan Committee, *North American Waterfowl Management Plan 2012*.

# 10

John F. Organ

# Private-Public Collaboration and Institutional Successes in North American Conservation

*The recovery and scientific management of continental North America's abundant wildlife resources stands as one of the great achievements in conservation history. Indeed, the global conservation narrative was significantly influenced by the bold experiment on this continent that has operated with distinction and measurable achievement for over a century. While the crux of this feat—the recovery and maintenance of abundant wildlife—is recognized as the North American Model's lasting achievement, the Model's success in this regard has depended upon multifaceted institutions that collectively support the conservation process. These institutions have, in turn, been supported by inventive private-public cooperation from very early in the Model's history. It is this organic complex that provides the inherent strength of the Model and has supported its longevity. This chapter focuses on some of these institutional components and their origins, as well as the roles they have played in and of themselves and as integrated processes within a complex whole. Each may be considered an individual success, including the emergence of the massive wildlife economy in support of conservation, the rise of academic programs and a professional elite, and the unparalleled diversity of the NGO conservation community.*

## Introduction

What is success? Government agencies, businesses, and other organizations rightfully strive to develop metrics before launching an initiative so progress toward stated goals can be measured.[1] What would the catalysts of the original conservation movement in North America—Theodore Roosevelt, George Bird Grinnell, Sir Wilfred Laurier, C. Gordon Hewitt—have set their sights on? What would the major innovators and thinkers of the wildlife conservation revolution during the Dust Bowl—Aldo Leopold, J. N. "Ding" Darling, Paul Errington, Hoyes Lloyd—have envisioned? What would the leaders of the environmental movement—Rachel Carson, Olaus Murie—have hoped for?

We can only glean from their writings what the pioneers of conservation might have truly envisioned as success. This is where context and scale are important. During the mid- to late nineteenth century, threats to wildlife were many and urgent. Roosevelt and Grinnell believed wildlife conservation was essential to maintenance of the character of the American people—a people who carved a society out of wilderness and were notable for self-reliance and a sense of fairness.[2] This long-term goal must have been tempered by the urgency brought on by urbanization and industrialization, the decline of big game species in the West, market hunting, and an absolute lack of governmental oversight of resource exploitation.[3] Successes came incrementally with reactionary efforts to deal with these urgent issues. Examples include the Yellowstone Park Protection Act (1894), the Lacey Act (1900), the Northwest Game Act

(1907), and the Migratory Bird Treaty (1916). Progress on other, more long-term initiatives, such as efforts to secure greater national and continental institutionalization of conservation through the 1908 Conference of Governors and the 1909 North American Conservation Congress, fell short. Yet progress did come from these incremental accomplishments, and with these successes came the legal and policy initiatives that collectively formed the foundation for what is now recognized as the North American Model of Wildlife Conservation.[4]

Herein, I focus not on the incremental successes spawned by a social movement on behalf of nature, unparalleled in the world, but rather on the creation and evolution of a dynamic institution, globally unique and enduring.

## Development of a Public-Private Conservation Institution

Overexploitation and decline of wildlife in North America during the nineteenth century have been well chronicled.[5] Private individuals and groups with vested interests, particularly hunters, but also wilderness advocates, initiated efforts to raise public awareness and halt unregulated take and other destructive uses of wildlife.[6] In what remains a remarkable demonstration of civic responsibility and grass roots democracy, private interests forced governments to take charge for the greater public good. The culmination and measurable success of these efforts was, in many instances, enactment of legislation at local, state, provincial, and national scales. In order for legislation to be enacted, however, governments had to have a legal, constitutional basis for exerting authority over wildlife. That authority is derived through government ownership of wildlife.

Ownership of wildlife in the United States and Canada is held in trust by state, provincial, and federal governments for the benefit of current and future generations of their citizens.[7] The common law basis for this in the United States is the 1842 Supreme Court decision in *Martin v. Waddell*, known as the Public Trust Doctrine.[8] The basis for the Public Trust Doctrine in the United States was English common law derived from the Magna Carta, which was directly applied in the American colonies prior to independence.[9] There is no comparable common law basis for public trust within Canada, but most provinces have statutory language prescribing trustee status that effectively establishes the same governmental oversight authority and responsibility.[10] Further, the legal basis for Canadian public trust in wildlife laws was affirmed as common to those in the United States by the Canadian Supreme Court in 2004 in *British Columbia v. Canadian Forest Products, Ltd.*, in which the court referred to the Institutes of Justinian (from which the Magna Carta was, in part, derived), English common law, and the evolution of the Public Trust Doctrine in the United States.[11]

Early conservation efforts were largely local and private, involving purchases of lands by wealthy individuals and groups of recreational hunters and hiring of wardens to exclude market hunters, such as on Carroll's Island in Maryland.[12] Many local efforts, such as those initiated by the New York Sportsmen's Club, went further and resulted in municipal or county-level ordinances.[13] While these were positive developments, the early leaders of the incipient conservation movement recognized that it was necessary for government to act in order to address the wider challenges facing the conservation of wildlife at the state and continental levels. The catalyst for national legislation was the battle to protect Yellowstone National Park from poachers. George Bird Grinnell, co-founder of the Boone and Crockett Club and editor of *Forest and Stream*, led a spirited battle on behalf of the club, which culminated in the Yellowstone Park Protection Act of 1874, the first federal anti-poaching law, sponsored by Congressman John Lacey of Iowa, who was also a member of the Boone and Crockett Club.[14] On its behalf, he subsequently sponsored the 1900 Game and Wild Birds Preservation and Disposition Act (also known as the Lacey Act), the most significant wildlife law enforcement act in United States history. The public-private conservation institution had become rooted.

By 1900 most states and provinces had formed wildlife management agencies, and a modest number of nongovernmental wildlife organizations and clubs existed.[15] These fledgling nongovernmental groups quickly helped foster conservation debate and enabled public opinion to more effectively engage and influence wildlife policy and legislation. Thus, for example, in the early twentieth century scientists and public activists began questioning the ecological basis for federal predator control programs in the United States.[16] This activism led, in some instances, to state laws restricting tools commonly used in animal control at the time.[17] In Canada, as well, hunting and fishing clubs played an integral role in the enforcement of game laws in remote areas, causing Theodore Roosevelt to note that wildlife thrived "to an extraordinary degree on these club lands."[18] Notably, Canada, a Dominion in the British Empire during the early twentieth century, largely rejected England's policies on wildlife conservation and moved in unison with the United States.[19] This cooperation was fostered by many private efforts, including those of the Boone and Crockett Club, an increasingly influential organization comprised of hunters from both countries. The great manifestation of this bi-national conservation approach was the passage of the extraordinary Migratory Bird Treaty of 1916 and the development of the associated flyway system of waterfowl management, in which governments, state and provincial agencies, and conservation organizations, many of them hunter based, from both nations, collaborate in science and policy.[20]

The conservation revolution that occurred in the 1930s further solidified the private sector's role in wildlife conservation. The 1930 American Game Policy ushered in a paradigm shift from conservative management of species through laws restricting take and better wildlife enforcement to a more diverse and proactive program that would strongly emphasize habitat management and wildlife restoration.[21] In the United States, passage of the Pittman-Robertson Wildlife Restoration Act of 1937 cemented relationships among hunters, the firearms industry, and state and federal wildlife agencies in that country. The law's taxation provisions effectively conjoined the vested interests of hunters wanting wildlife to pursue and of industry wanting customers and ultimately provided the financial fuel for government agencies to restore wildlife and habitats. Recruitment and training of professional wildlife biologists was further enhanced in the United States by the establishment in 1935 of the Cooperative Wildlife Research Unit Program, a public-private partnership in which industry, states, universities, and the federal government share in both direct funding and administration costs.[22]

Following World War II, the North American public became increasingly interested in wildlife issues, due in no small part to successful species and habitat restoration programs and the increased capacity of maturing wildlife agencies and institutions. In the United States, the number of nongovernmental wildlife organizations grew from fifty-six in 1945 to more than three hundred by the mid-1970s and more than four hundred by the 1980s.[23] The diversity of private-sector wildlife organizations shows patterns characteristic of social movements generally, such that growth leads to specialization and splintering along interest areas.[24] Thus the growth and diversity of wildlife groups eventually reflected special interests along species lines, as well as along political conservation agendas.[25]

This maturation of the wildlife conservation institution in the later twentieth century was also characterized by expanded legislative initiatives for broader environmental and conservation purposes. The dynamic between the private and public arms of the institution led to additional laws protecting endangered species, wetland habitats, clean air, clean water, and so on. Along with this maturation came stronger advocacy for wildlife policy from the private sector. While many nations do have wildlife agencies that take an authoritarian approach, the United States and Canada have moved from a more expert authority model to a more participatory one, in which the public has input into policy development, including, in many cases, actual trusteeship. Thus, governments are held accountable for their decisions by the public.[26]

The dynamic of this public-private conservation institution continues to evolve. The trend is toward greater participatory management and a broader suite of stakeholders.[27] This dynamic is not without its tensions, as evidenced by the multitude of direct-democracy ballot initiatives on specific conservation issues in the United States and the more general discontent over hunting and trapping issues expressed in many quarters in both countries.[28] Public debate, controversy, and litigation are ongoing within the institution.[29] Regardless, the fact that the United States and Canada have jurisdictional layers within government technical agencies to manage, protect, and conserve wildlife and the public has the means and access to influence policy formulation and implementation, as well as the means to challenge and hold government accountable, is an embedded and enduring success.

## Restoration of Species and Habitats

The transformation of the wildlife conservation enterprise in the 1930s from a program of restrictive use regulations toward active restoration of species and habitats was accomplished primarily through purposeful application of science to management.[30] A crowning achievement in the United States and Canada has been the development of scientific capacity within state, provincial, and federal management agencies, coupled with institutionalized collaboration with universities and other science entities.[31] Major policy initiatives, such as the United States Endangered Species Act of 1973, the Canadian Species at Risk Act of 2002, and the continental system of waterfowl management all have science embedded as a basis for decision making.[32]

The application of expanding scientific knowledge to wildlife management had other important implications. Such knowledge made it evident that wildlife populations would continue to decline if additional corrective measures were not taken.[33] Longer-term studies, especially, led to greater understanding of how factors such as habitat, climate, predation, and competition affected wildlife populations and allowed scientists to test and determine what effect management interventions could have.[34] More knowledge led to more questions, and as the questions became more challenging, wildlife science had to advance in order to address conservation needs. These advances pushed the boundaries of traditional wildlife management and led to advances in and applications of evolutionary biology, spatial ecology, landscape ecology, population biology, and quantitative science.[35]

The restoration of species, populations, and habitats has been well documented, with a focus during most of the twentieth century on hunted species.[36] Such game species were among the continent's first and most endangered wildlife. Market-driven overexploitation of such species and concurrent habitat loss garnered significant public concern that encouraged efforts for species recoveries.[37] As these hunted populations recovered and efforts to conserve habitats mounted, attention began to shift during the latter third of the twentieth century toward imperiled, nonhunted species. But could elements of law, policy, and principle comprising the North American Model of Wildlife Conservation provide a foundation for conserving rare and endangered wildlife species of all kinds?

Three significant pieces of wildlife protection legislation were passed in the United States, each superseding the other: the 1966 Endangered Species Preservation Act, the 1969 Endangered Species Conservation Act, and the 1973 Endangered Species Act (ESA).[38] In Canada, the Species at Risk Act (SARA), passed in 2002, established an independent agency of status review for wildlife in that country, the Committee on the Status of Endangered Wildlife in

Canada (COSEWIC). Both the ESA and SARA mandate the listing (designation under law) of species at risk of extinction, and both include provisions for protection, enforcement, and recovery measures for these listed species. In addition, the Fish and Wildlife Conservation Act of 1980 was passed in the United States to conserve nongame vertebrate species, including those *not* at risk of extinction. This act was not reauthorized, but in the year 2000, the Wildlife Conservation and Restoration Act, which had similar but broader provisions, was passed. The 2000 act received appropriated funding for only one year (2001), but Congress has annually provided funding under the Interior Appropriations Acts (State and Tribal Wildlife Grants). The foundation for these acts can be traced to the keystone principle of the Model: public ownership of wildlife and government's common law authority to hold wildlife in trust for the benefit of current and future generations.[39]

## Evolution of Conservation Agencies and Wildlife Science: An Achievement Providing Resilience

The cumulative addition of laws, policies, and principles in the nineteenth and twentieth centuries that comprise the North American Model enabled development of a robust public-private wildlife conservation institution, as described above. This institution was built within a culture that placed value on wildlife when alive and in public ownership, incorporated science in its management, and strongly supported sustainable use of wildlife, while also allowing wider public interests to participate in and benefit from its conservation efforts.

As mentioned earlier, public concern over environmental health led to expanded federal, state, and provincial laws to provide for clean air, clean water, and wilderness, as well as to reduce environmental contaminants and other external impacts to wildlife. The maturing wildlife conservation institution was a natural place for these programs to reside, and as a result, legislatures expanded programs and responsibilities for many federal, state, and provincial agencies.[40] In essence, the presence of an established, robust public-private wildlife conservation institution allowed its actors to organically form a nexus between wildlife conservation and general social environmental concerns. Resultant "superagencies" created expanded conservation opportunities and brought improved professional status in some instances, but freighted difficulties and governance challenges in others.[41] The advantages, in some jurisdictions, stemmed from the institutional expansion of the hunter's conservation ethic beyond fish and wildlife management to influence wider land-use decisions. In the United States, agency wildlife personnel were also elevated in their civil service status to a level on par with other government workers in certain jurisdictions. Previously, because of the independence of their agencies, most professional wildlife staff were not eligible for the rights of other government employees. Disadvantages stemmed from additional bureaucratic layers in decision-making and policy development as traditional wildlife agencies became one division within larger corporate structures or superagencies. This often resulted in less decision-making authority, less authority in allocation of funds, and greater politicization of decisions.[42] Regardless, the connection to broader environmental interests helped maintain many wildlife agencies' relevance in a changing society.

The scientific foundation of wildlife conservation began within the sole construct of biological science, but this too would expand over time. The strengthening public-private architecture of conservation, the breakdown of the expert authority governance model, and the rise of an expanded environmental ethic had gradually created a situation in which viewpoints from more diverse stakeholders, both internal and external to the conservation institution, could influence policy.[43] Nevertheless, as late as the 1930s, Aldo Leopold, one of North America's most influential conservation visionaries, could still observe how lack of or inappropriate engagement of stakeholders could stifle wildlife management.[44] This led him to observe:

> One of the anomalies of modern ecology is the creation of two groups, each of which seems barely aware of the existence of the other. The one studies the human community, almost as if it were a separate entity, and calls its findings sociology, economics and history. The other studies the plant and animal community and comfortably relegates the hodge-podge of politics to the liberal arts. The inevitable fusion of these two lines of thought will, perhaps, constitute the outstanding advance of this century.[45]

What Leopold envisioned would ultimately become known as the discipline of Human Dimensions of Wildlife Conservation.[46] The growth, expansion, and integration of social science applications in the wildlife conservation enterprise took place slowly, but since the 1970s has led to greater understanding of stakeholder needs and desires and consideration of the impacts of management decisions on them.[47] Integrating social science within wildlife conservation has made decisions affecting wildlife resources and their stakeholders more quantifiable and transparent.[48] Ultimately, science-based decision-making, a hallmark of the North American Model, has become more durable and democratic due to the integration of social and biological sciences. Using this knowledge matrix, challenges to science-based recommendations (often viewed as "expert authority") can be anticipated and addressed, greatly reducing the potential for legal or other civil disputes which are often counterproductive to conservation progress.[49]

Current and emerging modifications to wildlife conservation and its scientific basis have necessitated not only more quantitative approaches and integration of human dimensions, but also wider transdisciplinary integration to assess range-wide impacts to species from drought, fire, climate change, energy development, and agriculture. Wildlife biology is by its nature an integrative discipline,[50] and it has significantly embraced landscape ecology to address these types of wildlife conservation challenges. Doing so is helping provide policy-makers with probable future scenarios on which to base their decisions.[51] In these regards, the expansion and evolution of wildlife science exemplify an important attribute of the North American Model—its resilience, adaptability, and potential for growth and modification to address current and future conservation challenges. This flexibility in large part explains its resilience across a long period of social and ecological change.

The evolution and advancement of science through the public-private wildlife conservation institution are tremendous success stories in themselves, because they arose organically by way of public policy that holds wildlife in trust, sets limits on use and exploitation, fosters collaboration across borders, and gives citizens a role in conservation. The capacity for wildlife science to address questions with a broad geographic and political scope, the legal authority to conserve and protect wildlife, and the public's role in decision-making have coalesced toward what history may view as one of the greatest achievements in North American wildlife conservation: cooperative or pre-listing conservation. "Pre-listing conservation" refers to research and management designed to stem the decline of a wildlife population in order to prevent the need for placing it on a list of threatened, endangered, or at-risk species.[52]

## Cooperative Conservation: Manifestation of the Public-Private Coalition

Significant conflicts over wildlife conservation policy have arisen across both public and private sectors since the implementation of the ESA and SARA.[53] Certain stakeholders are opposed to restrictions on uses of their land that may impact species or are concerned with economic impacts related to restrictions on land use. Conversely, others are concerned that restrictions are inadequate to prevent destructive actions impacting species in question. In each instance, stakeholders believe their citizen privileges, rights, or core values are being violated, such as personal freedoms to enjoy nature in its fullest variety versus freedom to do whatever they desire on land they personally own. Thus, implementing public policy to conserve endangered and threatened species is one of the most challenging tasks faced by wildlife professionals, particularly where private lands are involved. Quite often, the ultimate decisions are made in the courts, not by wildlife professionals working in concert with stakeholders.[54] Wildlife professionals nevertheless must then implement the courts' orders.

In recent years, collaborative approaches known variously as "cooperative conservation" and "prelisting conservation" have been initiated.[55] These approaches entail engagement of affected stakeholders (often landowners and extractive industries) with scientists and wildlife managers from all jurisdictional levels. In a highly idealized portrayal, scientists provide evidence for management actions or alternative land management practices or mitigation efforts that might lead to species recovery; affected interests then agree to implement those strategies; and finally, wildlife managers factor these into their regulatory decisions. Throughout this process, there is interaction and communication among all parties, coupled with rigorous monitoring. The incentive for private interests to engage is to prevent a species listing and having to endure greater regulatory restrictions. The incentive for wildlife managers is to achieve voluntary cooperation in wildlife recovery and conservation efforts. For both groups there is a desire to avoid protracted and costly court proceedings.

Two major achievements in cooperative conservation occurred in the United States in 2015. These include decisions that listing the greater sage grouse (*Centrocercus urophasianus*) and the New England cottontail (*Sylvilagus transitionalis*) under the ESA were not warranted.[56] If these decisions endure and become a model for endangered species conservation, they will demonstrate to the world how a free society can continue to utilize abundant, publicly owned wildlife while restoring rare species through collaboration among government, industry, and private citizens.

## Conclusion

This chapter began by pondering what the catalysts of the original wildlife conservation movement in the United States and Canada would view as successes. In the late nineteenth and early twentieth centuries wildlife was indeed on the brink of disaster.[57] One can only imagine early conservationists' amazement at the present-day status of game populations and the acreage of public wildlife lands. The sheer number of nongovernment wildlife organizations would probably baffle them. The number of agencies and ranks of wildlife professionals would likely overwhelm them. The proliferation of science and its complexity would make their heads spin. Yet, one would imagine they would be impressed and heartened.

However, they may also be dismayed. They would likely be shocked by the massive human footprint on the land, as well as by the sheer numbers of people. They would likely be disturbed by the magnitude of both legal and illicit global trade in wildlife and their products. They may be concerned with the disparity in management attention given to various taxa. They might be saddened by sharp divides and polarization within the larger conservation community and yearn to see more focus on attaining common ground. They would definitely be upset to see that wildlife conservation is rarely more than an afterthought in most of today's political discussions. Leopold stated that we would not achieve true conservation until destructive uses of land were socially ostracized.[58] He certainly would be disappointed.

While the challenges facing wildlife conservation today and in the future might likely discourage our conservation ancestors, they just as likely would take pride in knowing that a social movement they began and nurtured had grown into an enduring institution, one composed of public agencies and private organizations and individuals, collaborating and contesting among themselves to achieve the greater good: namely, the conservation and sustainable management of wildlife for generations to come. The wildlife conservation institution is firmly rooted; only time will tell if it continues to branch and grow or senesce and be replaced.

*NOTES*

1. B. A. Radin, "The Government Performance and Results Act (GPRA): Hydra-Headed Monster or Flexible Management Tool," *Public Administrative Review* 58 (1998): 307–316.
2. H. W. Brands, *TR: The Last Romantic* (New York: Basic Books, 1997), 196–200; J. F. Organ, V. Geist, S. P. Mahoney, S. Williams, P. R. Krausman, G. R. Batcheller, T. A. Decker, R. Carmichael, P. Nanjappa, R. Regan, R. A. Medellin, R. Cantu, R. E. McCabe, S. Craven, G. M. Vecellio, and D. J. Decker, "The North American Model of Wildlife Conservation," *Wildlife Society Technical Review* 12, no. 4 (2012): 25.
3. S. A. Riess, *Sport in Industrial America, 1850–1920* (Wheeling, IL: Harlan Davidson, 1995), 4.
4. V. Geist, S. P. Mahoney, and J. F. Organ, "Why Hunting Has Defined the North American Model of Wildlife Conservation," *Transactions of the North American Wildlife and Natural Resources Conference* 66 (2001): 175–185; Organ et al., *The North American Model.*
5. J. B. Trefethen, *An American Crusade for Wildlife* (New York: Winchester, 1975), 3–65.
6. J. F. Reiger, *American Sportsmen and the Origins of Conservation* (New York: Winchester, 1975), 50.
7. V. Geist and J. F. Organ, "The Public Trust Foundation of the North American Model of Wildlife Conservation," *Northeast Wildlife* 58 (2004): 49–56.
8. Martin v. Waddell involved a landowner's claim that he owned oyster reefs in tidal waters of the Raritan River based on a grant to his ancestors by King Charles II in 1662 and had the right to exclude others from harvesting oysters. Chief Justice Roger Taney wrote in his decision, that has become known as the Public Trust Doctrine, that the King had no power to bestow such rights, since the Magna Carta had made such resources property to be held in trust by the King for the benefit of current and future generations. See J. L. Sax, "The Public Trust Doctrine in Natural Resource Law: Effective Judicial Intervention," *Michigan Law Review* 68, no. 3 (1970): 471–566. This is considered the definitive work on the origin and evolution of the Public Trust Doctrine; S. M. Horner, "Embryo, Not Fossil: Breathing Life into the Public Trust in Wildlife," *University of Wyoming College of Law, Land and Water Law Review* 35 (2000): 1–66.
9. Sax, "The Public Trust."
10. G. R. Batcheller, M. C. Bambery, L. Bies, and G. Roehm, "The Public Trust Doctrine: Implications for Wildlife Management and Conservation in the United States and Canada," *Wildlife Society Technical Review* 10-1 (2010).
11. In 1992, a fire destroyed a large area of forested land in British Columbia, for which the defendant logging company was largely responsible. At trial, the province was awarded damages for expenditures related to combating the fire and forest restoration, but was denied damages for the loss of revenue from harvestable trees and the loss of protected trees in environmentally sensitive areas. The Crown was unsuccessful in securing compensation for the environmental loss of harvestable and protected trees due to insufficient evidence and a failure to present a reliable means of measuring such loss. Although unsuccessful in this case, comments by the Supreme Court of Canada in its decision indicate that it has left the door open to future claims for this type of environmental damage. See British Columbia v. Canadian Forest Products Ltd., 2004 SCC 38; Sax, "The Public Trust"; M. C. Blumm and R. D. Guthrie, "Internationalizing the Public Trust Doctrine: Natural Law and Constitutional and Statutory Approaches to Fulfilling the Saxion Vision," *University of California Davis Law Review* 45 (2011): 741–808.
12. Trefethen, *An American Crusade*, 72.
13. J. F. Organ, "Fair Chase and Humane Treatment: Balancing the Ethics of Hunting and Trapping," *Transactions of the North American Wildlife and Natural Resources Conference* 63 (1998): 528–543.
14. *Forest and Stream* was a magazine founded in 1873 by George Hallock and purchased in 1879 by George Bird Grinnell. It was a powerful voice for conservation in its heyday. It was discontinued in 1930.
15. Reiger, *American Sportsmen*, 142.
16. T. R. Dunlap, *Saving America's Wildlife* (Princeton, NJ: Princeton University Press, 1988), 50.
17. L. Mighetto, *Wild Animals and American Environmental Ethics* (Tucson: University of Arizona Press, 1991), 58; Organ et al., "Balancing the Ethics."
18. C. G. Hewitt, *The Conservation of the Wild Life of Canada* (New York: Charles Scribner's Sons, 1921), 292.

19. Geist, Mahoney, and Organ, "Why Hunting Has Defined."
20. L. R. Jahn and C. Kabat, "Origin and Role," in *Flyways: Pioneering Waterfowl Management in North America*, ed. A. S. Hawkins, R. C. Hanson, H. K. Nelson, and H. M. Reeves (Washington, DC: United States Department of the Interior, Fish and Wildlife Service, 1984), 374–386; M. G. Anderson and P. I. Padding, "The North American Approach to Waterfowl Management: Synergy of Hunting and Habitat Conservation," *International Journal of Environmental Studies* 72, no. 5 (2015): 810–829.
21. A. Leopold, "Report to the American Game Conference on an American Game Policy," *Transactions of the Seventeenth American Game Conference* 17 (1930): 281–283.
22. W. R. Goforth, *The Cooperative Fish and Wildlife Research Unit Programs* (Washington, DC: US Department of the Interior, 1994).
23. Dunlap, *Saving America's Wildlife*, 109.
24. R. M. Muth, "Wildlife and Fisheries Policy at the Crossroads: Contemporary Sociocultural Values and Natural Resource Management," *Transactions of the Northeast Section of the Wildlife Society* 48 (1991): 170–174.
25. L. E. Baier, *Inside the Equal Access to Justice Act: Environmental Litigation and the Crippling Battle over America's Lands, Endangered Species, and Critical Habitats* (Lanham, MD: Rowman & Littlefield, 2016), 135.
26. R. B. Gill, "The Wildlife Professional Subculture: The Case of the Crazy Aunt," *Human Dimensions of Wildlife* 1, no. 1 (1996): 60–69; Organ et al., "Fair Chase and Humane Treatment."
27. C. A. Jacobson, J. F. Organ, D. J. Decker, G. R. Batcheller, and L. Carpenter, "A Conservation Institution for the 21st Century: Implications for State Wildlife Agencies," *Journal of Wildlife Management* 74, no. 2 (2010): 203–209.
28. S. J. Williamson, "Origins, History, and Current Use of Ballot Initiatives in Wildlife Management," *Human Dimensions of Wildlife* 3, no. 2 (2008): 51–59; M. A. Nie, "State Wildlife Policy and Management: The Scope and Bias of Political Conflict," *Public Administration Review* 64, no. 2 (2004): 221–233.
29. Baier, *Inside the Equal Access to Justice Act*, 271.
30. H. Kallman, ed., *Restoring America's Wildlife* (Washington, DC: US Department of the Interior, 1987), 59.
31. Wildlife Management Institute, *Organization, Authority and Programs of State Fish and Wildlife Agencies* (Washington, DC: Wildlife Management Institute, 1997).
32. National Academy of Science, *Science and the Endangered Species Act* (Washington, DC: National Academy Press, 1995); J. D. Nichols, M. C. Runge, F. A. Johnson, and B. K. Williams, "Adaptive Harvest Management of North American Waterfowl Populations: A Brief History and Future Prospects," *Journal of Ornithology* 148 (2007): S343–S349.
33. A. Leopold, *Report of a Game Survey of the North Central States* (Madison, WI: Sporting Arms and Ammunition Manufacturing Institute, 1931), 269; A. Leopold, *Game Management* (New York: Charles Scribner's Sons, 1933), 420.
34. P. L. Errington, "Some Contributions of a Fifteen-Year Local Study of the Northern Bobwhite to a Knowledge of Population Phenomena," *Ecological Monographs* 15, no. 1 (1945): 1–34.
35. V. Geist, *Mountain Sheep: A Study in Behavior and Evolution* (Chicago: University of Chicago Press, 1971); V. Geist, *Life Strategies, Human Evolution, and Environmental Design: Toward a Biological Theory of Health* (Berlin, Germany: Springer-Verlag, 1979); D. Tilman and P. Karieva, *Spatial Ecology: The Role of Space in Population Dynamics and Interspecific Interactions* (Princeton, NJ: Princeton University Press, 1975); R. T. T. Forman and M. Godron, *Landscape Ecology* (New York: Wiley and Sons, 1986); D. R. Anderson, "Optimal Exploitation Strategies for an Animal in a Markovian Environment: A Theory and an Example," *Ecology* 56, no. 6 (1975): 1281–1297; B. K. Williams, J. D. Nichols, and M. J. Conroy, *Analysis and Management of Animal Populations: Modeling, Estimation, and Decision-Making* (New York: Academic Press, 2001). For quantitative science, see K. P. Burnham and D. R. Anderson, *Model Selection and Interface: A Practical Information–Theoretic Approach* (Berlin, Germany: Springer-Verlag, 1998); K. McGarigal, S. Cushman, and S. Stafford, *Multivariate Statistics for Wildlife Ecology Research* (New York: Springer Science, 2000); N. T. Hobbs and M. B. Hoten, *Bayesian Models: A Statistical Primer for Ecologists* (Princeton, NJ: Princeton University Press, 2015).
36. Kallman, *Restoring America's Wildlife*, 31–57.
37. Organ et al., *The North American Model*; Trefethen, *An American Crusade*, 55–65.
38. M. J. Bean, *The Evolution of National Wildlife Law* (New York: Praeger, 1983).
39. V. Geist and J. F. Organ, "The Public Trust Foundation."
40. Wildlife Management Institute, *Organization, Authority and Programs*.
41. Wildlife Management Institute, *Organization, Authority and Programs*.

42. Wildlife Management Institute, *Organization, Authority and Programs*.
43. Gill, "The Wildlife Professional Subculture"; Organ et al., "Fair Chase and Humane Treatment."
44. C. Meine, *Aldo Leopold: His Life and Work* (Madison: University of Wisconsin Press, 1988), 359–360.
45. C. Meine, *Aldo Leopold*, 359–360.
46. D. J. Decker, S. J. Riley, and W. F. Siemer, *Human Dimensions of Wildlife Management* (Baltimore: Johns Hopkins University Press, 2012).
47. S. J. Riley, D. J. Decker, L. H. Carpenter, J. F. Organ, W. F. Siemer, G. F. Mattfield, and G. Parsons, "The Essence of Wildlife Management," *Wildlife Society Bulletin* 30, no. 2 (2002): 585–593.
48. M. J. Conroy and J. T. Peterson, *Decision Making in Natural Resource Management: A Structured Adaptive Approach* (Oxford, UK: Wiley-Blackwell, 2013), 4.
49. Organ et al., "Fair Chase and Humane Treatment"; Williamson, "Origins, History and Current Use."
50. J. F. Organ, "The Wildlife Professional," in *Wildlife Conservation and Management: Contemporary Principles and Practices*, ed. P. R. Krausman and J. W. Cain, III (Baltimore: Johns Hopkins University Press, 2013), 24–33.
51. O. Bastian, "Landscape Ecology—Towards a Unified Discipline?" *Landscape Ecology* 16, no. 8 (2001): 757–766; J. A. Bissonette, *Wildlife and Landscape Ecology: Effects of Pattern and Scale* (New York: Springer-Verlag, 1997), 3; T. C. Edwards, G. G. Moisen, T. S. Frescino, and J. J. Lawler, "Modeling Multiple Ecological Scales to Link Landscape Theory to Wildlife Conservation," in *Landscape Ecology and Resource Management: Linking Theory with Practice*, ed. J. A. Bissonette and I. Storch (Washington, DC: Island Press, 2003), 153–175; J. Wu, "Landscape Ecology, Cross-Disciplinarity, and Sustainability Science," *Landscape Ecology* 21, no. 1 (2006): 1–4.
52. Baier, *Inside the Equal Access to Justice Act*; C. J. Donlan, ed., *Proactive Strategies for Protecting Species* (Berkeley: University of California Press, 2015), 5.
53. Baier, *Inside the Equal Access to Justice Act*.
54. Baier, *Inside the Equal Access to Justice Act*.
55. Donlan, *Proactive Strategies*, 147–218.
56. US Fish and Wildlife Service, Endangered and Threatened Wildlife and Plants; 12-Month Finding on a Petition to List Greater Sage-Grouse (*Centrocercus urophasianus*) as an Endangered Species, Proposed Rule (2015), Federal Register 80: 59858–59942; US Fish and Wildlife Service, Endangered and Threatened Wildlife and Plants; 12-Month Finding on a Petition to List the New England Cottontail as an Endangered or Threatened Species, Proposed Rule (2015), Federal Register 80: 55286–55304.
57. Reiger, *American Sportsmen*, 27; Trefethen, *An American Crusade*, 55–65.
58. A. Leopold, "Wildlife in American Culture," *Journal of Wildlife Management* 7, no. 1 (1943): 1–6.

11

Leonard A. Brennan,
David G. Hewitt, and
Shane P. Mahoney

# Social, Economic, and Ecological Challenges to the North American Model of Wildlife Conservation

*The diversity and abundance of wildlife on this continent is a testament to the success of the North American Model of Wildlife Conservation. The Model, however, and indeed North American conservation, is facing an uncertain future. Its fate rests to a large degree not only on its ability to adapt to a range of social, ecological, and economic challenges, but also on the capacity of its proponents to address criticism regarding the validity and the exclusivity of certain components of the Model when applied to contemporary wildlife issues. In this chapter, we highlight current challenges to the Model (and to conservation, generally). This discussion includes global conservation challenges (e.g., population increase, climate change, economics, commercialization and privatization of land, funding, and lack of education), as well as criticism and conflicts posed in current literature.*

## Introduction

As we move forward through the twenty-first century, there is no doubt that numerous elements of our changing world challenge various tenets of the North American Model of Wildlife Conservation. These challenges are diverse and complex, and they are related to substantive changes in our domestic and international cultures and economies, as well as to shortcomings inherent in the Model itself.

Mahoney and colleagues have identified seven significant challenges—listed below—that threaten the North American Model of Wildlife Conservation today.[1] They emphasized, however, that these seven factors represent "*some but certainly not all of the most important challenges facing conservation practice in North America.*"[2]

1. Increasing human population
2. Globalization
3. Urbanization and the human-nature divide
4. Novel ecosystems
5. Lack of connectivity (or fragmentation)
6. Funding, fragility, and resilience
7. Abundance and superabundance

Our goal here is to elaborate on these seven challenges, but also expand beyond them to explore both modern and historic realities that pose difficulties for the Model as currently structured. The discussion, which is not exhaustive, is intended to provide insight into the range of challenges this century-old approach to wildlife rescue and conservation now confronts. We conclude with some recommendations that may be used to modify the North American Model and effectively deal with these challenges.

Challenges to conservation *writ large* and to the sustainable use of wildlife, specifically, are, de facto, challenges to the Model itself. Furthermore, while

North American conservation approaches include both active management (e.g., regulated hunting) and protectionist philosophies (e.g., national parks), the history and functioning of conservation practice in Canada and the United States remain strongly influenced by utilitarian values and by those publics engaged in the sustainable harvest of wild animals.[3] As a consequence, the escalating debate over the relevance and validity, as well as the social acceptance, of hunting (especially) and angling have direct and significant consequence to the North American Model and will be presented as such in our discussions here.

Challenges to the Model involve far more than threats to the sustainable use of wildlife. Many threats actually arise from within the hunting community itself. This is a critical point as it reflects the potential for a successful system of conservation to be undermined not only by conflicts arising from broad social changes and evolving public attitudes, which are in themselves inevitable realities, but also by historically influential and supportive agencies proving themselves incapable or unwilling to respond appropriately to evolving social norms and expectations. The latter is, by contrast, not an inevitability at all. It is simply a failing—a choice of positions that can undermine the Model and wildlife conservation in the long run.

Thus, we emphasize throughout this review our perception that North American conservation, generally, and the Model, specifically, are challenged both from without and from within. The former refers to large scale processes such as climate change and escalating habitat loss while the latter include inherent, historic weaknesses of the Model, as well as processes directed toward such things as wildlife privatization and genetic manipulation of wild animals. Some of these challenges are recent and more acute, while others are more long term and include historic inadequacies, such as the Model's failure to build a less frictive, more cooperative approach to achieving conservation success. While none of these issues individually represents a critical threat to the Model's future, collectively they do imply that some change in approach to North American conservation is both inevitable and required.

## An Uninformed Public

Conservationists have done a regrettably poor job at explaining conservation realities to the wider public in North America. For example, there is very little attention paid to conservation issues or challenges in elementary or secondary schools in Canada or the United States. In higher education circles, the only students exposed to conservation issues are, in general, those who choose a career in renewable natural resources or environmental science, or take some kind of wildlife or conservation-related class as a nonmajor to round out a general education requirement. Yet building an informed public is critical to conservation awareness and engagement, especially given that society's values and norms change over time. If the North American Model is to maintain relevance and gain broader support, it is essential that the public understands how and why this approach to conservation exists and how and why it has been successful. This requires much greater emphasis on conservation issues throughout basic educational programs, as well as more effective communication of wildlife issues in general.

## Prioritization of "Game" Species and Consumptive Users

While wildlife management policies should, ideally, apply to all taxa, the dedicated funding and specific stakeholder and advocacy base traditionally dominated by consumptive users of wildlife has likely distorted this comprehensive approach in Canada and the United States.[4] It is not surprising, therefore, that sustainable wildlife management in North America can appear to sectors of the wider public as primarily self-serving, such that efforts preferentially target exploited species because they are exploited. An objective overview of this perspective is critical for assessing the Model's ongoing relevance to wildlife conservation and for defining its inherent limitations for effectively communicating with the wider public. The ongoing decrease in numbers of consumptive users of wildlife in both countries brings emphasis and urgency to this challenge.

## An Incomplete Historical Narrative

The Model betrays its own history when it fails to recognize the formative contributions made by women and leaders of the nonconsumptive, more preservationist side of the conservation movement.[5] The lack of recognition given to the visionary efforts of wilderness advocates such as John Muir to establish national parks and the wider protected areas movement, for example, is readily apparent. Similarly, the role of women in ending the commercial market for bird feathers and in founding the Audubon Society is rarely mentioned, though their efforts were certainly recognized by conservation leaders of the day.[6] To its detriment, the traditional narrative also fails to address how prevalent social inequalities between genders, races, and classes during the development of the Model negatively impacted its practical inclusivity.[7] These narrative vacancies may, in consequence, limit incentives for individuals belonging to such user groups to engage in current as well as future wildlife conservation discussions.

## First Nations Traditions and Rights

Failure to effectively incorporate First Nations peoples and their traditions, perspectives, and rights within the conservation system of North America is a historic weakness of the Model and one with enduring implications. This legacy of exclusion has helped fuel tensions in both countries with regard to wildlife access, use, and harvest regulation. The actual, as well as perceived, discrepancies between First Nations' versus other publics' access to and rights of harvest of wild resources are significant as they challenge the principle of equality of access to wildlife and can, thus, undermine public support for the Model in general. These tensions need to be addressed.

## An Increasing Human Population

The global human population is projected to surge to 8.5 billion by 2030, an increase of more than 1 billion people in just fifteen years.[8] In North America, as in other regions, the rising human population and consequent escalation in natural resource consumption pose a considerable challenge to wildlife conservation efforts. As human populations grow, productive lands are developed for settlement and commercial or industrial use, reducing their availability and utility to wildlife. Wildlife are often killed or displaced by such land-use alteration or otherwise impacted by outcomes and processes including pollution and landscape fragmentation, which in turn lead to permanent losses in biodiversity and ecosystem resilience.[9]

Furthermore, as human populations expand and natural areas decline, human-wildlife conflicts, such as crop depredation, motor vehicle collisions, and disease transfers, increase. Such conflicts, in turn, demand policy responses, which often polarize public attitudes. As a result, tolerance for wildlife abundance, as well as for sustainable use practices such as hunting and angling, can decrease.[10] Thus if solutions to such conflicts are not considered adequate, local support for conservation efforts, and for the Model, specifically, is likely to decline.[11]

## Globalization

Globalization is the process by which ideas and organizations develop international influence or start operating on an international scale. For the nations of the world and their various conservation agendas, this means that distant publics can influence local and regional policies through effective lobbying and political and economic pressure. North

America is not immune from such influence, and important aspects of the Model, such as regulated wildlife harvests, are subject to scrutiny not only by the citizens of Canada and the United States but also by publics, NGOs, and policy institutions around the world. Global perspectives derived from the ongoing evolution of ethical, cultural, and moral valuations will pose an increasing challenge to the North American Model over time, both from within the region and from without.

Living within a global community has major implications for all conservation approaches. Input by international stakeholders will create a progressively more varied mix of attitudes and expectations for wildlife conservation and management, and this will add complexity to policy-making. It is also likely to introduce novel concepts to traditional North American conservation practices.[12]

## Urbanization and the Human-Nature Divide

Industrial development and technological innovation have not just enabled higher density human populations in North America, but have also resulted in the reduction of economic opportunities in rural areas, encouraging, often through economic necessity, increasing urbanization in Canada and the United States.[13] The United Nations projects that 68 percent of the global population will live in cities by 2050, an increase of 13 percent over the current (2018) statistic, whereas 82 percent of North Americans already live in urban areas.[14] The large-scale displacement of local and rural culture, traditionally linked to more direct human engagement with nature, has highly meaningful implications for wildlife conservation and the North American Model. As landscapes dominated by natural diversity are replaced with landscapes dominated by technodiversity, human behaviors and awareness of nature and values toward it change.

Increasing human remoteness from nature, from wilderness, and from the origins of food results in a reduced familiarity with and sense of value associated with natural resources, including wildlife.[15] Sustainable-use traditions, for example, which are inherent prescriptions for the Model, are gradually lost as traditional knowledge of engagement with nature is depleted. This loss of knowledge often creates obstacles to maintaining public support for practical conservation approaches and hands-on wildlife management.[16] As values and attitudes concerning living resources shift from utilitarian to intrinsic, anti-use sentiment is likely to weaken general acceptance of and support for many aspects of the North American Model.[17]

## Novel Ecosystems

Humans have altered vast areas of the planet to the extent that many biogeographers now consider Earth to be experiencing a new geological epoch, the Anthropocene era. Extensive urban, suburban, and industrialized landscapes now exist on at least half to three quarters of the globe. The presence of these unprecedented technoecosystems, driven by high inputs of energy from carbon, nuclear, and environmental (wind and solar) sources, represents unique challenges for conservation scientists and policy-makers, which are not adequately addressed by the Model as it is currently structured.

For example, large-scale environmental impacts resulting from human activity, such as climate change, can cause entirely new suites of habitat-organism complexes to develop.[18] Exotic and invasive species colonizing new regions through direct or accidental human intervention are significantly affecting local ecosystem compositions. In the near future, conservationists will need to know how to best manage these novel ecosystems. However, the speed and complexity of these changes to natural architecture will inevitably impede knowledge gains sufficient to the task.[19] Policy-makers will be challenged to make informed decisions quickly enough to prevent further biodiversity loss, while at the same time being faced with the increasing presence of novel species and significantly altered distributions of many common ones.

## Climate Change

Of all the ecological factors that pose challenges to the Model, the ecological effects of climate change on wildlife populations will likely be the most profound. Climate change is an especially troubling challenge because its effects are uncertain and are likely to vary among species and habitats, from highly negative to highly positive. Climate change is likely to exacerbate problems for rare species with specific habitat requirements but potentially improve conditions for invasive generalist species, such as feral pigs. Some of these changes will have unexpected consequences. In North America, the shift of temperate species northward has already influenced Inuit dialect, with Indigenous people borrowing from the English language to name and describe new resident species.[20]

As habitats and species' distributions change, wildlife managers will be faced with surveying and managing novel assemblages of wildlife. Thus wildlife management resources that are already scarce are likely to be spread even thinner. The uncertainty of impacts will complicate planning and dilute the effectiveness of science, an important underpinning of the North American Model, in wildlife management decisions, at least in the near term.

## Abundance and Superabundance

While there are the many instances of wildlife depopulation in the wake of human development, some species of native wildlife thrive in the human-impacted landscapes common across much of North America. White-tailed deer, elk, snow geese and Canada geese, raccoons, beavers, and even black bears are often highly abundant in such areas, resulting in increased human-wildlife conflict.[21] Some abundant or superabundant species, especially herbivores, cause ecological damage, thereby threatening other species of flora and fauna.[22] Additionally, overly abundant species can present various challenges to the Model by influencing public attitudes, primarily toward common and problematic wildlife species. Public enthusiasm for wildlife, and therefore support for its thoughtful management, is often reduced when the public experiences negative engagement with abundant or superabundant species. For example, goose feces on lawns, deer eating decorative plants and causing motor vehicle accidents, birds striking airplanes, and beavers cutting down prized trees and flooding yards all cause people to devalue wildlife.[23] Indeed, simply seeing the same species everyday can make those species less exciting and devalued in the public consciousness. The implication here is that when wildlife's value declines, the availability of resources to support wildlife management often declines accordingly.

Deer, raccoons, opossums, beavers, urban grackles, Canada geese, and invasive exotics (rats, pigs, starlings) are examples of free-ranging animals that have for a long time been perceived to negatively impact the lives of many citizens. For example, for more than a century agricultural producers have complained about coyotes, wolves, deer, elk, and various species of black birds that eat grain crops, livestock, and other valued species. But there are also examples of wild species that have more recently caused such problems and have, therefore, drifted from treasured wildlife to ecological troublemakers, such as snow geese, elk (in national parks), red fox (in some limited circumstances), barred owls (at least relative to spotted owls), feral pigs, and others.

Moving forward, scientists and wildlife managers will be challenged with evaluating the appropriate abundance of species in human dominated systems. This will be a difficult task, since modern society holds a very complicated and somewhat contradictory set of expectations for wildlife conservation efforts.[24] The Model emerged at a time when many of these challenges either did not exist or were far less intense both because of drastically reduced wildlife abundance and because there were far fewer people and industrialized landscapes.

## Connectivity

Human activities and developments fragment wildlife habitat and reduce connectivity among habitat mosaics. Decreasing connectivity, or inversely, increasing fragmentation, effectively disrupts movements of wild animals and thus compromises dispersal and migration, which are essential factors for population persistence and conservation. Connectivity represents an especially acute problem for migratory and wide-ranging species.

In a recent study of the effects of habitat fragmentation on northern bobwhite quail populations in Texas, Oklahoma, and Louisiana over the past forty years, Miller and colleagues found an incredibly sensitive threshold relationship between bobwhite population declines and the presence of impervious surfaces related to human infrastructure development.[25] In this study, Breeding Bird Survey Routes that had accumulated about 0.7 percent of impervious surfaces over forty years also had the greatest declines in bobwhite numbers. Impervious surfaces are not limited just to roads; they also include sidewalks, parking lots, buildings, airport tarmacs, and any other human construction that does not allow rainfall to percolate directly into soil, but rather requires runoff to other areas.

While we seldom have such long-term and refined data, connectivity is obviously a critical factor for all wildlife species, including those, like quail, that are hunted. Reductions in huntable wildlife populations due to fragmentation or other human-induced habitat changes inevitably dovetail with, and undermine, supportive constituencies and associated financial support for the Model. Connectivity is also strongly related to issues of private land ownership and conservation. More than half of the land in the United States is privately owned, and extensive public land in Canada is controlled corporately through leases for resource extraction.[26] Efforts to improve habitat connectivity, therefore, cannot be successful without considering privately owned and privately controlled land.

## Declining Participation in Hunting

Hunting has been a central component of the Model for more than a century. With the exception of First Nations traditions, however, hunting in North America is now almost wholly a recreational activity. As such, hunting competes with a myriad of other recreational activities for peoples' attention and time. Hunting is also a relatively expensive recreational activity, not only for tangibles like clothing and equipment, but also in travel and time.

In addition, broad cultural and social changes are impacting the acceptability of numerous aspects of North America's hunting traditions, the hunting of carnivores being just one example. Thus, for various reasons, the proportion of people who hunt in North America is declining (to less than 5 percent), and the hunting public is ageing. The average age of a North American hunter is now greater than forty-five years, and recruitment of new hunters is insufficient to prevent further declines in participation. It is highly likely that the Model will be faced with an increasingly reduced constituency of consumptive users of wildlife going forward.

In contrast, many nonconsumptive outdoor activities such as birding, wildlife photography, kayaking, and biking are increasing in popularity. While these trends pose definite challenges, they also present an opportunity to expand social and political support for a modified and more inclusive Model.

## Access to Land and Hunting Opportunity

Lack of access to places to hunt is considered a major factor driving the decline in hunting participation in North America referred to above, and in the United States there are few places where this is more true than in Texas. Because it is largely a "private lands state," landowners in Texas do many things to directly promote wildlife conservation, as do many nongovernmental organizations such as Ducks Unlimited, the Quail Coalition, and the Coastal Conservation Association. These organizations, among others, receive millions of dollars of support annually from private landowners. Moreover, some of the largest private ranches in the world are found in Texas, and most of these actively promote hunting either for family and friends or for the wider public through leasing arrangements.

However, unless you own a large tract of land or are close friends with someone who does, obtaining access to hunt in Texas can mean you need money. A lot of money. For example, in South Texas, typical hunting lease fees run about US$15.00 per acre per year. While at first glance this seems reasonable, the size of most South Texas hunting leases is greater than one thousand acres, which would require a fee of US$15,000 per year. Often leases range upward to more than fifty thousand acres, requiring an annual payment of US$750,000. Relatively few hunters have US$15,000 in disposable income per year to allocate to this recreational activity, much less three quarters of a million. Yet, that is the economic reality facing individuals wishing to access private lands for hunting in Texas. And the situation is not restricted to that state, but applies wherever private land exists. It does not matter that populations of wild animals—game animals and otherwise—are held in public trust in both Canada and the United States. If you control access to the land, de facto you control access to wildlife and to hunting opportunities as well. With more than 60 percent of all lands in the United States in private hands, this is an issue of great importance to conservation generally and to hunting's future specifically.[27]

## The Special Case of Public Lands

Public lands, especially public lands in the American West and across Northern Canada, provide access for millions of hunters and other outdoor recreationists in North America. While overharvest of species such as deer and elk can be problematic in some places, lands held in public ownership generally represent one of the last remaining resources freely available to people interested in hunting and related outdoor pursuits. It is a troubling development for conservationists and hunters that some legislators, especially at the federal level in the United States, are hostile to continued federal stewardship of this public trust resource. The idea of ceding such lands to the states—which, for the most part, do not have the necessary resources to adequately manage them—or selling them outright should be of concern to all citizens, whether they are interested in wildlife conservation or not. Loss of these lands for hunting would have a devastating impact on the viability and resilience of the North American Model.

The relatively recent development and rapid expansion of fracking for oil and gas extraction on public and private lands present new challenges to US wildlife conservation efforts. Add to this the proliferation of wind farms, solar panel fields, and mountain top removal for coal extraction, and you have a trifecta of energy development threats to wildlife, and, by extension, to the North American Model. Restoration efforts for grassland, shrubland, and early successional habitats for small game birds and mammals, as well as other desirable nongame species, are key to mitigating at least some of the losses of wildlife habitat—and access to wildlife habitat for hunting—from these activities.

## Inadequate Funding

Historically, hunters and anglers have been the primary economic stakeholders of the Model, especially in the United States and increasingly in Canada, with their direct economic contributions deriving from special taxation on equipment and through fees for hunting permits. Although the broader public certainly benefits from the Model and its wildlife conservation successes, the majority of North Americans do not seem to be aware of the benefits the Model provides, nor how it is paid for and functions. This is true even of politicians, who should be aware of how these funding mechanisms work to support wildlife conservation, but often are not.

Indeed, it is not uncommon for state legislators in the United States to attempt to divert hunter- and angler-derived direct taxation (Pittman-Robertson and/or Dingel-Johnson) funds from state wildlife resource agencies to other political priorities, even though this is inappropriate from a legal perspective.[28] More often than not, some kind of intervention from the federal sector is required to make state legislatures aware of the legal consequences of such interventions. Political problems like these would likely be diminished if the Model enjoyed financial support from a broader public coalition. For decades, various conservation organizations have tried to encourage legislators in the United States and Canada to broaden funding mechanisms in support of conservation, but to no avail. In current political climates where new taxes are often considered economically destructive, meaningfully expanding the funding base for conservation continues to be a massive challenge.

Yet, additional funding is definitely needed, not just to meet increasing future requirements and address new threats to North American conservation efforts, but to support current practices and efforts. Without adequate funding, it will not be possible to successfully protect imperiled or endangered species; nor will it be possible to conserve and restore the resilient ecosystems that maintain wild biodiversity and ecosystem services.[29]

## Human Relationships with Predators

Even while understanding our evolutionary relationships as omnivores and the essential role predation plays in natural communities, the issue of human relationships with carnivores is an especially complicated challenge for the Model. Humans have evolved strong reactions to carnivores that remain part of our psyche today. Even with the emergence of what might be called a more empathetic attitude toward carnivores from society generally and within some factions of the conservation community, it is impossible to eliminate all sense of fear when hiking in grizzly bear country, or somehow having a feeling of being stalked by a mountain lion in areas where those animals are known to exist. This is a fundamental element of the human condition that is impossible to escape. At the same time, we hold a deep fascination for the power and beauty of carnivores. These conflicting attitudes mean that how hunters engage with carnivores will be especially significant from a public perception point of view and how institutions within the Model work to manage carnivore populations will be of long-term significance for ensuring continued public support.

Today, however, many hunters—and many other people—still believe that elimination of upper trophic level carnivores, especially wolves, will inevitably result in more deer, more elk, and better hunting opportunities. This perspective also holds true among many upland game bird hunters, who believe that reduction of raptors and mesomammal predators such as raccoons, skunks, opossums, and the like will result is greater quail abundance, even if the science fails to support such efforts. To the broader public, the idea of killing one group of animals to improve opportunities for hunting another is often considered backward, if not reprehensible, and thus espousing such views can reflect poorly on hunters in general. Hunters (and the Model) may be better served by viewing the world as an ecologist does, rather than as a privileged predator.

## Wildlife as Reservoirs for Disease

Related both to issues of wildlife abundance and to the complicated relationship between humans and predators, the fear that wildlife can transmit diseases to humans, livestock, and pets is a growing challenge that must be overcome if the Model is to persist. In the past, anthrax, leprosy, and plague decimated millions of people. Today, various tick-borne human illnesses such as Lyme disease, which have indirect life cycles associated with small and large mammals, are increasing. Therefore, it is not surprising that some people have developed negative views of wildlife and see them as potential threats to public health. Increasingly, wildlife may also be a source of pathogens that threaten the livestock industry. White-tailed deer, elk, and bison have been implicated in sustaining or spreading bovine tuberculosis, brucellosis, and babesiosis. Feral pigs have introduced pathogens to agricultural crops, resulting in human food poisoning and produce recalls. Perhaps the greatest wildlife disease–related threat to the Model, however, is the emergence and spread of the prion-based chronic wasting disease. This disease, first characterized in captive breeding facilities for deer, has the capacity to fundamentally undermine wildlife management efforts in North America, drastically alter public perceptions of wildlife as a source of high-quality food, and create massive uncertainty with respect to wildlife's capacity to serve as a reservoir for human pathogens. In a worst-case scenario it could transform wildlife from something highly desirable to the category of vermin in the public mind. The impacts for conservation and the Model from such a scenario could be catastrophic.

## Breeding and Privatization of Captive Wildlife: The Case of White-Tailed Deer in Texas

The recent confirmation that chronic wasting disease (CWD) is now present in white-tailed deer in Texas is a classic case history of how social, economic, and ecological factors can combine to create an acute challenge to the Model and of how the captive breeding of wildlife is leading to major challenges for North American conservation.

White-tailed deer are culturally important in Texas and have enormous economic value to the state. Texas has a long history of deer hunting, the activity is broadly accepted there, and over 600,000 people hunt deer in the state every year. Interest in deer hunting has resulted in a desire to increase the number and size of deer that can be harvested, which, in turn, has led to a strong interest in deer management and huge investments in those efforts. For example, in 2012 deer managers in Texas spent US$3 million surveying deer from helicopters and at least US$100 million in supplemental feed and bait for deer. Because 95 percent of land in Texas is private, much of that deer management is done by private landowners.

Despite a free-ranging population of 3.5–4.7 million deer, Texas has developed a huge captive deer industry in which the animals are held in small paddocks to be selectively bred for large antlers as incentives for high-paying hunters. In 2012, there were 1,300 permitted captive deer facilities in Texas containing over 100,000 deer. Under this system, sires and dams are carefully selected for breeding, and artificial insemination is common. Captive-raised deer are sold to other deer breeders who wish to obtain offspring selected for large antlers to incorporate into the deer herd on their property or for release directly into fenced pastures for hunting. This business model requires a high monetary investment in facilities and personnel on the part of the deer breeder. The deer are vaccinated and treated with antibiotics and other medications. The business model also requires frequent movement of deer among captive facilities and from breeding facilities to hunting pastures or ranches. Deer in these captive facilities exhibit traits unlike those of native deer, such as exceedingly large antlers, rapid growth rates, dependence on artificial feed, and little fear of people.

So how does a captive deer management and manipulation program like that in Texas threaten the Model? Fundamentally, challenges arise because regulations by state wildlife agencies, implemented to manage wildlife as a public trust resource, impede the business model of captive cervid facilities. In response, the captive deer industry then seeks to change regulatory oversight of captive deer from the Parks and Wildlife Department to the Department of Agriculture in Texas and in other states and provinces where the captive deer industry exists. Such a change would allow captive deer to be managed as livestock and therefore regulatory decisions would no longer need to consider populations of free-ranging deer (or other wildlife) that are not of interest to the hunting and wider public. While such changes are opposed vigorously by many wildlife enthusiasts across the country, nine American states currently place sole regulatory oversight of captive deer facilities with their state departments of agriculture.[30] Some captive deer operators wish to further limit industry oversight by having captive deer classified as private property and therefore not subject to oversight by any governmental agency, a move entirely against one of the founding principles of wildlife conservation in North America and of the Model that ownership of wildlife is a public trust.

These efforts to erode regulatory oversight of the captive cervid industry are important for two main reasons. First, wildlife management decisions will no longer be made in light of the public's interest in healthy, productive wildlife populations but instead will be determined by the economic self-interest of the captive facility's owner or the captive deer industry. This change in motivation is a problem for the Model, because, as in the days of market hunting, the actions of a small number of people who benefit financially can negatively impact wildlife resources for everyone. An example of such an action is the captive deer industry's challenge to regulations designed to reduce the spread of CWD by restricting transport of captive deer. Such regulations disrupt the captive deer industry's business model, but are necessary to prevent disease spread and protect wild deer populations.

A second reason a change in oversight and ownership of wildlife is a threat to the Model is that under public trust management, all citizens have an interest in wildlife because the wildlife is, essentially, theirs. Wildlife matter to the North American public as evidenced by the 86 million US residents who participate in wildlife-watching activities each year and the hundreds of millions of dollars donated to wildlife-based organizations.[31] Making captive deer private property will reduce the motivation for people to support wildlife conservation. Just as it would be difficult to generate funds from the public to support farmer Joe's herd of cattle, so it would be difficult to motivate the public to support and care about wildlife if it is held as private property.

Another socially based threat to the Model from captive deer facilities arises from the objective of these facilities to produce deer with exceptionally large antlers for hunting. Revenue from hunting provides the majority of money for wildlife management in the United States, yet only 5 percent of the public hunts.[32] Hunting remains a viable recreational activity because 74 percent of the public supports it.[33] However, public support for hunting drops to low levels if the perceived motivation of the hunter is to acquire a trophy and if the animal pursued does not have a reasonable opportunity to escape.[34] The public is likely to find distasteful a hunting enterprise that focuses on providing trophy status quarry raised in captivity and then released into a fenced paddock to be shot. Protecting the privilege to hunt and maintaining public support are crucial to the current model of wildlife conservation. Captive deer operations manifestly threaten both.

But why, given these realities, has the captive deer industry developed in Texas and other areas? Economic changes are one reason. While the number of big game hunters in Texas grew by 160,000 between 1996 and 2011, areas available for deer hunting declined as urban, suburban, industrial, and agricultural development consumed ever larger areas where deer occurred and were accessible.[35] As hunting opportunities declined and affluence increased, hunters were willing and able to pay more to hunt. This increased willingness to pay more for hunting access resulted in land costs being influenced by a property's recreational potential, lands being leased at higher values, and hunters' increasing willingness to pay vast sums to shoot large-antlered deer.[36] In a positive feedback process, then, the amount of money hunters are willing to pay to hunt deer has, in turn, enabled the captive deer industry to become economically viable.

The interplay between these social factors and deer biology has further provided an incentive to the captive deer industry. Smaller land-holdings and more fragmented deer populations combined with the species' demographic rates have made it difficult for some wildlife managers to meet their management objectives for free-ranging deer numbers on their property. They seek to overcome these limitations by increasing the intensity of their management, including through the establishment of captive cervid facilities, which thereby leads to further growth in the industry.

The challenges to the North American Model posed by the captive deer industry are broad and far-reaching. Wildlife conservation in North America became possible only once markets for wildlife were eliminated. The captive deer industry has essentially reestablished markets for deer. Deer are sold to the highest bidder often for hundreds of thousands of dollars. Deer management decisions become heavily influenced by the markets and what is expedient for the deer breeder, with little thought given to the impacts on deer populations or public support for hunting and wildlife management. Hostility on the part of the captive deer industry toward regulations meant to control CWD is a perfect example of the clear and direct threat wildlife markets can have on both wildlife management policies and the Model.

In addition to the wider economic implications from a loss in public support for hunting and the revenues this activity generates for conservation, state wildlife agencies incur significant additional costs in their efforts to oversee and regulate captive deer facilities. Although this spending is necessary to meet the agency's oversight obligations, the resources used for it have to be diverted from other wildlife management activities.

## How Might the Model Change or Adapt?

It is clear from the material presented here that myriad challenges to the Model must be addressed if it is to adapt and survive through the next fifty to one hundred years. The declining numbers of sustainable-use advocates, hunters especially, will inevitably lead to a weakening of the North American conservation approach, at least in terms of how it has been funded and how it has functioned for the last one hundred years. At the same time, differing interests and engagements with wildlife are expanding. For example, expenditures in the United States on wildlife viewing grew from US$71 to US$86 million in the last ten years, while, overall, people spent US$76 billion per year pursuing nonconsumptive wildlife interests.[37] We argue that both explicit and implicit support, financial and otherwise, from these nonconsumptive wildlife and fisheries–related activities will be critical for the Model to adapt in our rapidly changing world. This will also mean that state, provincial, and federal conservation institutions will have to be more responsive to this community of interests and reflect this in their programs, staffing, and policies.[38] Some of the efforts these institutions should engage are listed below.

### Public Education

Public education certainly has the potential to inform people about conservation realities, but it has to be done correctly. Take for example, the highly successful Smokey Bear campaign promoted by resource agencies and the media over many decades to warn about the risks of forest fires in the United States and Canada. Even though the message of the benefits of managed forests and the use of prescribed fire—as opposed to the damages caused by wild fires—was lost, the Smokey Bear phenomenon shows that it is possible to make the public aware of resource conservation issues. It further demonstrates that it is possible to move the public from awareness to active support for solutions, which in this case generated political capital that resulted in virtually unlimited financial resources for fire suppression. In hindsight, we might have wished for a less extreme focus on fire prevention and more on regulated fire and more active forest management. The point is, however, conservation education can work—which makes it critical that we get the messages right!

## Environmental Education in Elementary and Secondary Schools

Increasing the environmental literacy of young people is essential to the future of the Model in particular and conservation in general. Children have an innate, natural curiosity about nature and wild animals; they just need opportunities to learn about wildlife and spend time outdoors. State wildlife agencies and the private sector recognize the importance of engaging young people in outdoor activity, and there are many effective programs across the continent focused on this goal. For example, the Texas Wildlife Association supports a variety of educational programs using age-appropriate activities to engage children from kindergarten through high school.[39] In 2016, these programs engaged over 500,000 students in that state with events ranging from video conferences and webinars to wildlife necropsies and weeklong intensive wildlife-based summer camps.

In addition to these broad outdoor educational activities, many of which also cover the importance of hunting in wildlife conservation, there is the more focused Texas Youth Hunting Program, which is a collaborative endeavor between Texas Parks and Wildlife Department and the Texas Wildlife Association.[40] The Texas Youth Hunting Program provides high-quality hunts for hundreds of youth every year, ranging from upland game birds and waterfowl to javelina and deer. Every hunt is led by a trained hunt-master and requires that each youth have a sponsoring adult with him or her. There are discussions on hunter ethics, hunting safety, and the role of hunting in wildlife conservation. Youth hunters are also given guidance on processing the animals they harvest and take the meat from their animal home to cook and enjoy.

These efforts, and many others like them now ongoing in various states and provinces, need to be encouraged and specifically designed to build broad coalitions of interest in the outdoors, which include but are not restricted to hunting and angling. The future of the Model, as we have already alluded, rests in a wider framework of interest and support than currently exists, and the sustainable use of wild resources must find its place within this wider matrix, rather than stand apart from it.

## Bridging the Disconnect between Hunters and Nonhunters

Hunting activity remains a vital element within the North American Model, and maintaining public support for this activity is critical to its future. It is probably impossible and certainly very difficult to bridge the cultural disconnect between hunters and antihunters, and therefore we focus here on how to encourage shared values between hunters and nonhunters in modern society. We define "nonhunters" as people who do not hunt, but who are not opposed to the activity. Today, for example, even in the wildlife profession itself, there are many wildlife scientists and managers who do not hunt or fish, but who are not opposed to other people doing so.

Like the disconnect that often prevails between management and research components of wildlife conservation, the disconnect between hunters and nonhunters is rooted in a complex array of social and cultural factors and is not always oppositional.[41] Cultural values and norms and family upbringing, as well as many other personal and economic priorities are the most common reasons why many people may not participate in hunting activity themselves, and yet remain unopposed to it or even be strongly supportive of it.

Encouraging such relationships and promoting the positive values of hunting will help maintain public support for it. Such values include the personal health benefits derived from a wholesome outdoor activity that requires physical exertion and involves direct interaction with the natural world. Hunting can foster close ties with family and friends through time spent together outdoors, and, as described above, hunting provides access to healthy and sustainable sources of food. It also imparts a clear understanding of where our nourishment comes from and of the value of wildlife habitat as the basis for one aspect of our food security.

Hunting also provides broader social and ecological benefits such as control of high-density wildlife populations that would otherwise damage ecosystems or come into conflict with people in various ways, such as through agricultural or property damage, motor-vehicle collisions, or disease transfer. Hunting-related expenditures bring many economic benefits and frequently help transfer wealth from cities to economically struggling rural communities, which are closer to where the hunting actually takes place. Furthermore, the license fees and excise tax money generated by hunters provide resources to manage not only huntable species but all wildlife, though certainly too few in the general public know or appreciate this point.

Clearly, then, there is a basis on which a shared dialogue between hunters and the broader public can be constructed. This shared dialogue will prove critical in maintaining the viability and relevance of the North American Model. Hunters and anglers form obvious minorities in North American society today, and their lifestyle choices are of increasing interest to the general public and, for certain sectors, of increasing concern. Public support is a nonnegotiable requirement for the continued existence of sustainable use of wild living resources in North America and for the Model as we know it today.

## Game Meat and Locavores

A menu at an upscale restaurant in Fredericksburg, Texas, lists nilgai antelope steak at US$55.00 per eight-ounce serving (US$110.00 per pound): this is one example of an expanding trend in food consumption patterns in North American society. Consumers are increasingly aware of the benefits of eating foods produced more traditionally and near their home and of the potential health risks, real or imagined, associated with foods produced by industrial agriculture. In this context, game meat is considered an appealing alternative that is healthy and sustainable. In terms of the Model and its future support, various programs such as Michigan State University's Gourmet Gone Wild, and Conservation Visions Inc.'s Wild Harvest Initiative® have tapped into this trend. These efforts are communicating the health and food security benefits of eating wild meat and could well assist in recruiting new hunters, thereby helping to build support for hunting within the wider public and building wider coalitions in support of wildlife and habitat management. Such initiatives may help offset, to some extent, the demographic trends now leading to a decline in hunter numbers.

## Increasing Access to Private Lands

Maintaining access to private lands for hunting and angling is important if the North American Model is to thrive in the future. Yet this is a highly problematic issue. Increasingly, in many areas of North America landowners are reluctant to allow hunters access to their lands and often not because of their opposition to hunting per se but because they are fearful of exposure to liability. To deal with such circumstances and help ensure hunting access to private lands, numerous approaches have been developed. For example, there are now insurance policies available for hunters and landowners that protect the landowner from such legal action. There are also state regulations that provide some liability protection, such as a Texas statute that limits landowner liability with respect to recreational activities taking place on their land.[42]

In addition, many states now have programs that use hunter dollars to provide access for hunting on private lands enrolled in specifically designed programs. For example, the Texas Parks and Wildlife Department has a program that pays owners of croplands and pastures to allow the public to shoot doves during fall and winter. In Montana, 1,324 landowners enrolled nearly 7.5 million acres in the state's Block Management program, which allows hunters free access to these private lands. Despite such efforts, however, maintaining access to private lands for hunting remains a serious challenge to the Model's future and further innovations are required.

## Conservation Easements and Other Private Land Conservation Incentives

A conservation easement is a legal agreement whereby a landowner "donates" or forfeits certain aspects of his or her rights to develop a piece of land for personal or economic gain. In return for this donation of development rights, the landowner reduces the overall economic value of that land, and thus obtains a lower burden of annual taxes, and if the parcel of land is sufficiently large, inheritance taxes as well. The development rights included in a conservation easement agreement are then typically donated to a nonprofit organization, such as The Nature Conservancy, for example, and are to be held in such trust in perpetuity. Conservation easements are a creative and important conservation tool because they use market-driven mechanisms to promote conservation of wild spaces, and they do so for private lands. While the conservation practices these easements specify can vary, they often do not exclude hunting and angling access as part of the easement conditions and can, in fact, specifically allow such activities.

The economic impact of such easements can be very attractive to landowners. In Florida, for example, privately owned parcels of land less than forty acres can be 100 percent exempt from property taxes if they are under conservation easement, and parcels of land larger than forty acres can receive up to a 50 percent reduction in annual taxes. These are strong incentives for conservation initiatives that protect wildlife habitat and traditional land uses in a state where the human population is rapidly increasing and demands for land development are intense.

In one example from Florida, suburban development was closing in on the southern end of a hunting-permitted property consisting of about 2,500 acres of land, which, excluding the house and outbuildings, was assessed at a market value of about US$8 million. By putting this property in a conservation easement, the owner was granted a 50 percent reduction in annual property taxes. During the past twenty-five years, more than 100,000 acres of additional neighboring private lands have also been protected from development under similar conservation easements. Clearly, reducing private tax burdens in such a way can be a critical tactic for protecting wildlife habitat first and foremost and, where specified, protecting hunting and angling and other traditional wildlife related activities on private lands.

In Texas, tax policies for agricultural land allow for property valuations far lower than would be the case for similar-sized residential or other property types. Traditionally, however, agricultural exemptions usually require a landowner to plant crops or graze livestock on the land in question. A relatively recent wildlife exemption policy implemented in Texas now allows for property valuations at the reduced agricultural exemption rate, but does not require the landowner to plant crops or graze livestock. Instead, all the landowner has to do is develop a land and wildlife management plan under advisement from a qualified wildlife professional and file that document with the county tax assessor. This is one more policy example in which tax incentives to private citizens can pay big dividends for wildlife and, in a prohunting state like Texas, for sustainable wildlife use and, ultimately, for North America's conservation model.

## Keeping Public Lands Public

We are naïve to think that public lands in Canada and the United States will necessarily remain public in perpetuity. As noted previously, some political factions have engaged in an aggressive assault on public land ownership in the United States. Many of these advocates want to "return" such lands to state ownership, which is oxymoronic because either (a) such lands never belonged to the states or (b) placing such lands in federal public trust was a condition of statehood in the first place. Other individuals simply want to sell such lands outright to the highest bidder, moving them directly to private ownership.

Either of these proposals carry great risk for North American conservation and the Model, as the likely outcomes would be less recreational access to the public, for hunters and nonhunters alike. This, in turn, would likely lead over the long term to reduced interest and support for wildlife conservation and the potential loss of large areas of wildlife habitat due to development opportunities. While conservation incentives involving private lands are critically important, keeping these public lands in public ownership should also be a high priority in support of the North American Model.

## Broad-Based Funding

Establishing a broader funding base in support of the Model should be a political priority, and we suggest it might be wise to start with a small, discrete initiative such as an excise tax on camouflage clothing. Although there are articles of camouflage clothing that would never make it into a deer blind—surely more than one high school student has worn a camouflage tuxedo jacket to a prom—our assumption is that the vast majority of such clothing is worn by hunters. And even if it is not, so what? A huge amount of ammunition is used on target ranges for recreational and competitive shooting sports and not for hunting purposes, yet through the Pittman-Robertson Act tax sales of this ammunition directly help to support the Model and wildlife conservation. Why not do something as simple as a small excise tax on camouflage as well?

Ultimately, this camouflage funding model could be extended to all outdoor clothing, hiking boots, bicycles, field guides, tents, coolers, binoculars, and spotting scopes—at a minimum. Linking an excise tax on such items with a statement on the price tag or store receipt to the effect that a percentage of the purchase price will be used to support wildlife and natural resource conservation would also provide a de facto conservation message with every purchase. Just as United States' arms and ammunition manufactures recognized in the 1930s that robust wildlife populations were necessary for their business model and thus helped fund conservation, industries that today rely on nonconsumptive interactions with wildlife should rally to support similar taxation in support of what is, actually, the basis of their business.

## Overcoming Ecological Challenges

The ecological factors that represent challenges to the Model are perhaps the most difficult to overcome. Reversing the impact of the human footprint on biodiversity, ecosystem resilience, and ecological processes is a tall order. So is dealing with more specific large-scale problems such as climate change. Being an environmental manager is much easier than being an environmental engineer, as we have come to realize. Nevertheless, what is clear is that we need to quickly develop ideas for how to adapt our conservation model to deal with these new realities, while at the same time seeking solutions to them in the longer term. One thing is for certain: we will have to increasingly find ways to work proactively for wildlife in urban and industrialized areas because such landscapes will be an increasing part of the context within which the North American Model plays out.

## Conclusion

Clearly, there are both challenges and threats to the North American Model of Wildlife Conservation. As we have noted, these challenges and threats are widespread, diverse, and pernicious, although they also vary in the extent to which they have the potential to impact the Model. Nevertheless, the current situation clearly points to a need for proactive measures that can be used to "retool" the Model as we move forward into the next century of conservation efforts. While some suggestions offered here and elsewhere may seem radical, they provide a means for us to start the critical conversations we need in order to make serious inroads to overcoming the threats and challenges to the North America's highly successful conservation approach.[43] How effective we are in this task will significantly influence the future of wildlife on this continent.

*NOTES*

1. S. P. Mahoney and D. Cobb, "Future Challenges to the Model," *Wildlife Professional* 4 (2010): 83–85; S. P. Mahoney, ed., *Conservation and Hunting in North America*, monograph issue, *International Journal of Environmental Studies* 72, no. 5 (2015): 731–899.
2. Mahoney, *Conservation and Hunting*; emphasis added.
3. J. R. Heffelfinger, V. Geist, and W. Wishart, "The Role of Hunting in North American Wildlife Conservation," *International Journal of Environmental Studies* 70, no. 3 (2013): 399–413; S. P. Mahoney and J. J. Jackson, "Enshrining Hunting as a Foundation for Conservation—The North American Model," *International Journal of Environmental Studies* 70, no. 3 (2013): 448–459.
4. M. J. Peterson and M. P. Nelson, "Why the North American Model of Wildlife Conservation Is Problematic for Modern Wildlife Management," *Human Dimensions of Wildlife* 22 (2016): 43–54.
5. Peterson and Nelson, "Why the North American Model"; M. P. Nelson, J. A. Vucetich, P. C. Paquet, and J. K. Bump, "An Inadequate Construct? North American Model: What's Flawed, What's Missing, What's Needed," *Wildlife Professional* 5 (Summer 2013): 58–60.
6. Peterson and Nelson, "Why the North American Model."
7. Andrea M. Feldpausch-Parker, Israel D. Parker, and Elizabeth S. Vidon, "Privileging Consumptive Use: A Critique of Ideology, Power, and Discourse in the North American Model of Wildlife Conservation," *Conservation and Society* 15, no. 1 (2017): 33–40.
8. United Nations, "Global Issues, Population," November 9, 2016, http://www.un.org/en/sections/issues-depth/population/.
9. S. P. Mahoney, P. Krausman, and J. N. Weir, "Challenges for Conservation and Sustainable Use in North America," *International Journal of Environmental Studies* 72, no. 5 (2015): 880–881.
10. Mahoney, Krausman, and Weir, "Challenges," 881.
11. World Wildlife Fund, "Human-Wildlife Conflict," World Wildlife Fund, Knowledge Hub, May 13, 2018, https://wwf.panda.org/our_work/wildlife/problems/human_animal_conflict/.
12. Mahoney, Krausman, and Weir, "Challenges," 881–882.
13. Mahoney, Krausman, and Weir, "Challenges," 882.
14. United Nations, "68% of the World Population Projected to Live in Urban Areas by 2050, Says UN," *United Nations Department of Economic and Social Affairs, News*, May 16, 2018, https://www.un.org/development/desa/en/news/population/2018-revision-of-world-urbanization-prospects.html.
15. Mahoney, Krausman, and Weir, "Challenges," 882.
16. Mahoney, Krausman, and Weir, "Challenges," 882.
17. Nelson et al., "An Inadequate Construct?"
18. Mahoney, Krausman, and Weir, "Challenges," 882.
19. Mahoney, Krausman, and Weir, "Challenges," 882.
20. Mahoney, Krausman, and Weir, "Challenges," 883.
21. J. Sterba, *Nature Wars: The Incredible Story of How Wildlife Comebacks Turned Backyards into Battlegrounds* (New York: Crown Publishers, 2012).
22. C. D. Ankey, "An Embarrassment of Riches: Too Many Geese," *Journal of Wildlife Management* 60, no. 2 (1996): 217–223; S. D. Cote, "Impacts on Ecosystems," in *Biology and Management of White-Tailed Deer*, ed. D. G. Hewitt (Boca Raton, FL: CRC Press, 2011).
23. M. R. Conover, *Resolving Human-Wildlife Conflicts: The Science of Wildlife Damage Management* (Boca Raton, FL: CRC Press, 2001).
24. Mahoney, Krausman, and Weir, "Challenges," 885.
25. K. S. Miller, L. A. Brennan, H. L. Perotto-Baldivieso, F. Hernandez, E. Grahmann, A. Z. Okay, X. B. Wu, M. J. Peterson, H. Hannusch, J. Mara and J. Robles, "Impacts of Habitat Fragmentation on Northern Bobwhites in the Gulf Coast Prairie Landscape Conservation Cooperative," *National Quail Symposium Proceedings* 8 (2017), https://trace.tennessee.edu/nqsp/vol8/iss1/40.
26. Mahoney, Krausman, and Weir, "Challenges," 883–884.
27. S. M. Stern, "Encouraging Conservation on Private Lands: A Behavioral Analysis of Financial Incentives," *Arizona Law Review* 48 (2006): 541–583.
28. The Federal Aid in Wildlife Restoration Act (16 U.S.C. 669–669i; 50 Stat. 917) of September 2, 1937, is commonly called the "Pittman-Robertson Act." It has been amended several times and provides federal aid to states for management and restoration of wildlife. Funds from an 11-percent excise tax on sporting arms

and ammunition (Internal Revenue Code of 1954, sec. 4161(b)) are appropriated to the Secretary of the Interior and apportioned to states on a formula basis for paying up to 75 percent of the cost approved projects. Project activities include acquisition and improvement of wildlife habitat, introduction of wildlife into suitable habitat, research into wildlife problems, surveys and inventories of wildlife problems, acquisition and development of access facilities for public use, and hunter education programs, including construction and operation of public target ranges. (Digest of Federal Resource Laws of Interest to the US Fish and Wildlife Service, "Federal Aid in Wildlife Restoration Act," May 27, 2015, https://www.fws.gov/laws/lawsdigest/fawild.html.)

The Federal Aid in Sport Fish Restoration Act (16 U.S.C. 777–777k, 64 Stat. 430) was established on August 9, 1950, and has been amended several times. It is commonly called the Dingell-Johnson Act, or Wallop-Breaux Act, and provides federal aid to the states for management and restoration of fish having "material value in connection with sport or recreation in the marine and/or fresh waters of the United States." In addition, amendments to the act provide funds to the states for aquatic education, wetlands restoration, boat safety and clean vessel sanitation devices (pumpouts), and a nontrailerable boat program.

Funds distributed to states for the various programs funded in the act are collected in an account known as the Sport Fish Restoration Account, one of two accounts in the Aquatic Resources Trust Fund established under the authority of the internal revenue code (26 U.S.C. 9504(a)). Unless otherwise specified in the act, funds are permanently appropriated (see P.L. 136, August 31, 1951; 65 Stat. 262). Funds are derived from a 10-percent excise tax on certain items of sport fishing tackle (Internal Revenue Code of 1954, sec. 4161), a 3-percent excise tax on fish finders and electric trolling motors, import duties on fishing tackle, yachts and pleasure craft, interest on the account, and a portion of motorboat fuel tax revenues and small engine fuel taxes authorized under the Internal Revenue Code (Sec. 9503). To be eligible to participate in the Federal Aid in Sport Fish Restoration program, states are required to assent to this law and pass laws for the conservation of fish, which includes a prohibition against the diversion of license fees for any other purpose than the administration of the state fish department.

Funds for the permanently appropriated states sport fish program are apportioned on a formula basis for paying up to 75 percent of the cost of approved projects, which includes acquisition and improvement of sport fish habitat, stocking of fish, research into fishery resource problems, surveys and inventories of sport fish populations, and acquisition and development of access facilities for public use. Funds for the remaining programs under the act must be authorized to be appropriated from the Sport Fish Restoration Account by Congress. (Digest of Federal Resource Laws of Interest to the US Fish and Wildlife Service, "Federal Aid in Sport Fish Restoration Act," May 27, 2015, https://www.fws.gov/laws/lawsdigest/fasport.html.)

29. Mahoney, Krausman, and Weir, "Challenges," 884.
30. Michigan Department of Natural Resources, "Chronic Wasting Disease and Cervidae Regulations in North America," August 21, 2017, http://cwd-info.org/wp-content/uploads/2017/08/CWDRegstableState-Province.pdf.
31. US Department of the Interior, US Fish and Wildlife Service, and US Department of Commerce, US Census Bureau, "2016 National Survey of Fishing, Hunting, and Wildlife-Associated Recreation," April 24, 2018, https://wsfrprograms.fws.gov/Subpages/NationalSurvey/nat_survey2016.pdf.
32. US Department of the Interior et al. "2016 National Survey."
33. National Shooting Sports Foundation, *Americans' Attitudes toward Hunting, Fishing, and Target Shooting 2011* (Newtown, CT: National Shooting Sports Foundation, 2011).
34. D. J. Decker, R. C. Stedman, L. R. Larson, and W. F. Siemer, "Hunting for Wildlife Management in America," *Wildlife Professional* 9, no. 1 (2015): 26–29.
35. US Department of the Interior, US Fish and Wildlife Service, and US Department of Commerce, US Census Bureau, "1996 National Survey of Fishing, Hunting, and Wildlife-Associated Recreation," September 7, 2018, https://www.census.gov/prod/3/97pubs/fhw96nat.pdf; US Department of the Interior, US Fish and Wildlife Service, and US Department of Commerce, US Census Bureau, "1996 National Survey of Fishing, Hunting, and Wildlife-Associated Recreation," September 7, 2018, https://www.census.gov/prod/2012pubs/fhw11-nat.pdf.
36. J. Baen, "The Growing Importance and Value Implications of Recreational Hunting Leases to Agricultural Land Investors," *Journal of Real Estate Research* 14, no. 3 (1997): 399–414.
37. US Department of the Interior et al., "2016 National Survey."
38. S. G. Clark and M. B. Rutherford, "The North American Model of Wildlife Conservation: An Analysis of Challenges and Adaptive Options," in *Large Carnivore Conservation: Integrating Science and Policy in the American*

*West*, ed. S. G. Clark and M. B. Rutherford (Chicago, Illinois: University of Chicago Press, 2014), 289–324.

39. Texas Wildlife Association, "Program Areas, Youth Education," April 9, 2012, https://www.texas-wildlife.org/program-areas/category/youth/.
40. Texas Youth Hunting Program, "Youth Hunters," June 23, 2014, http://tyhp.alamark.com/youth-hunters/.
41. J. P. Sands, S. J. DeMaso, M. J. Shnupp, and L. A. Brennan, eds., *Wildlife Science: Connecting Research with Management* (Boca Raton, FL: CRC Press, 2012).
42. Texas Constitution and Statutes, "Civil Practice and Remedies Code, Title 4. Liability in Tort, Chapter 75. Limitation of Landowners' Liability," September 1, 2017, https://statutes.capitol.texas.gov/Docs/CP/htm/CP.75.htm.
43. S. Wagner, "Is It Time to Rethink the North American Model of Wildlife Conservation?" *Outdoor Life*, January 28, 2016, https://www.outdoorlife.com/articles/hunting/2016/01/it-time-rethink-north-american-model-wildlife-conservation?dom=odl.

# 12 A Comparison of the North American Model to Other Conservation Approaches

Rosie Cooney

*The North American Model is based on the Public Trust Doctrine, which ensures democratic access to wildlife resources. Historically, wealth, status, and land ownership were not the basis for access to, or use of, wildlife on the continent; rather, public use was the means by which wildlife was recovered and managed. Different approaches have emerged around the world where private land ownership and other models of social organization have supported conservation. One of the great achievements and surprising aspects of North American conservation, especially given that excessive exploitation had led to serious declines in wildlife populations, was the recognition that self-interest must play a role in any long-term, effective strategy.*

## Introduction

The intent of this chapter is to set the North American Model within global context—to highlight and discuss some aspects of the Model in comparison to approaches to conserving and managing wildlife that have been taken in other parts of the globe, both developed and developing. The North American system is unique—and differs sharply from comparable systems elsewhere. My focus here is primarily on the larger terrestrial animals that are generally the most heavily used by people for food, clothing, recreation, and adornment.

The Model is described here as North American in a conceptual, rather than geographical context, in reference to the practices of Canada and the United States. Wildlife conservation and management in Mexico, the Caribbean, and Central America have pursued very different trajectories shaped by their various political, social, and cultural dynamics.[1] And of course wildlife management in North America did not begin with "the North American Model"—Indigenous custodianship of wildlife predates it by many thousands of years.

Comparative analysis of wildlife management systems and approaches is a markedly under-researched area. Very few systematic analyses exist. This chapter does not therefore attempt to be comprehensive, the more so because fully describing even any one country's policy and management system for wild species is itself a major endeavor. It is intended rather to give a sense of how some key features of the Model echo or contrast with approaches taken elsewhere.

## The Success of the North American Model in a Global Context

Before beginning this examination, however, it is useful to ask how the Model is faring in terms of its success in conserving wildlife. Clearly it has achieved enormous success in recovering many larger vertebrate species over the last hundred or so years from a point of acute, unregulated over-hunting and wide-scale depletion to one of healthy or abundant population.[2] While comparing its achievements with other management systems is intrinsically challenging—countries all face different ecological, demographic, governance, and development problems—there are few places in the world that can boast such a track record.

Southern Africa—South Africa and Namibia most markedly—have seen impressive increases in many wildlife populations and the land area devoted to conservation over the last three to four decades,[3] including iconic elephants and rhinos. Europe is also seeing a recovery of wildlife including bears, wolves, and lynx, albeit at a more limited scale and starting considerably more recently.[4] In Russia, during the Soviet regime, many wildlife populations recovered under tighter hunting regulation,[5] although it is unclear whether these successes have endured the breakup of the Soviet Union. Other countries can boast more limited successes. Australia, for example, has recovered saltwater crocodiles from severe depletion due to unregulated hunting up until the 1970s to a level now thought to be at carrying capacity.[6] During this century India has succeeded in reversing an established downward trend in tiger and elephant populations,[7] while Nepal has successfully recovered its rhino population in recent years after it was heavily poached in the country's civil war.

Globally, however, wildlife populations (roughly corresponding to terrestrial vertebrates) continue to dramatically decline—with current populations (in terms of numbers of individuals) being less than half of what they were in 1970.[8] The Model stands out against this backdrop as indeed one of the most impressive success stories of global conservation.

But humans, politics, ecosystems, and social histories are complex and diverse, and there is no single right way to conserve wildlife populations while continuing to draw on them for the multiple tangible and intangible benefits that humans gain from their presence and use. I turn now to contrasting the Model to some other models that have emerged around the world.

Here I highlight three features of the Model for particular exploration. The first is the Public Trust Doctrine—that the government holds wildlife in trust for the citizenry as a whole. The second is the role of hunting in the management model. The third is the proscription on commercial use of (dead) wildlife—meaning, with limited exceptions, no sale of wildlife products.

## Who Owns Wildlife?

The question of ownership of wildlife is both interesting and of great practical significance for wildlife conservation and management. "Ownership" is often not a particularly helpful term—conceptually, ownership is composed of a "bundle of rights," but these may not all be held by a single "owner." Ownership rights can include rights to access, use, and benefit from and sell a given product or experience; for wildlife these can be translated into rights to hunt and manage wildlife and to derive benefits from hunting, tourism, and sale of wildlife products. Under wildlife management systems governments often exercise some of these rights while private landholders or recreational users may exercise others. In the Model, according to the Public Trust Doctrine, wildlife is owned by no one but is held in trust by government for the benefit of all citizens.[9] Governments then hold and exercise most ownership rights but must do so on behalf of the people and in their interests.

Government ownership of wildlife is a common approach globally, remaining the case, for example, across most African jurisdictions even where there have been decades of emphasis on devolution of wildlife management to communities.[10] It is specified in law in countries such as Tajikistan, China, and Australia.[11] In part, the dominance of this system may be because wildlife is a "fugitive resource": animals are (unless fenced in) no respecters of property or tenure boundaries and may move freely across the borders of private land or protected areas, making private ownership problematic. This is exacerbated in the case of migratory resources.

However, other approaches exist. In South Africa, for example, wildlife are formally *res nullius*—no one owns them.[12] However, on fenced properties private landowners effectively exercise rights of ownership, being empowered to manage, use, control access to, and sell wildlife. In Namibia, private landholder and communal conservancies exercise most ownership rights over most wildlife, with certain restrictions imposed by government on use of some species.[13] In many Western European countries, ownership of wildlife is vested in private landowners, although this may (as in some cases in the United Kingdom) or may not (as in France) automatically confer the right to hunt.[14] This preponderance of landowner control of wildlife in Europe reflects a long history of elite control of the use of and benefits from hunting of certain game species—a history that contrasts sharply with the typical colonial and postcolonial model of centralized state control.

Ownership rights are important because they determine who makes decisions concerning wildlife, who can benefit from it, and, less tangibly, who feels a sense of responsibility and stewardship for it. The distilled experience of decades of more or less successful attempts at community-based wildlife conservation, primarily in Africa, has emphasized the importance of ownership (or more accurately, rights associated with ownership) in determining conservation outcomes.[15] Inadequate devolution of ownership rights from the state has often limited empowerment of communities to manage their resources in ways that meet their needs, stymied any sense of stewardship of wildlife, and curtailed the benefits (and incentives for conservation) that communities can gain from sustainable use. Central governments are typically reluctant to cede control of wildlife (and corresponding benefit streams) to private landowners or communities. Namibia provides the clearest example of strong private/community ownership rights and how powerful they can be in incentivizing conservation. Under this model, private landholders or community conservancies can under certain conditions gain effective ownership of wildlife and thereby make most decisions concerning it and being entitled to retain 100 percent of commercial benefits gained through sustainable use.[16] The conservation benefits of this national approach are well known, with dramatic expansions of wildlife populations and land area dedicated to wildlife and reduced poaching and human-wildlife conflict.

Given this basis of experience, it has become practically a truism in some circles that the key problem undermining wildlife conservation efforts in many parts of the world is the reluctance of central governments to empower local communities with rights to manage and benefit from wildlife. But in the Model, apart from lands where First Nations hold recognized rights and claims, most of the key decisions concerning wildlife management are taken by governments. The people and communities who live with wildlife participate in management to a very limited extent. While there is a level of community consultation and input, this is a long way from the sort of devolved community management or, failing that, comanagement that is called for across much of the developing world. Why then does the Model work and not fall into the traps outlined above? Why doesn't government control in North America lead to alienation of people from wildlife conservation, resentment of conservation authorities, and inadequate incentives to motivate conservation?

While I don't propose to attempt to fully answer this question, part of the answer is no doubt the Public Trust Doctrine—that while ownership of wildlife is legally vested in the government ownership and control, authorities are obliged to exercise their responsibilities on behalf of the public. Governance is sufficiently robust that this obligation is duly discharged, in a way that would be unlikely in many countries.[17] Despite lack of ownership or comanagement rights, the people who live with wildlife receive benefits from it in various ways, and conflict is adequately ameliorated. Government subsidies in both the United States and Canada enable conservation successes without imposing economic or other practical hardships on citizens.

A further part of the answer is likely to be the economic setting. While human-wildlife conflict is real in North America, it pales by contrast with the costs suffered by rural communities in some other parts of the world. Elephants kill people in India alone at the rate of more than one a day; crocodiles in Africa kill over a hundred people a year.[18] A single elephant incursion can destroy an entire year's food store for a family. Children can be killed on the way to school or fetching water. These costs imposed by wildlife are many orders of magnitude greater and qualitatively different in the existential threat they pose in parts of the developing world than in the United States or Canada. The corresponding incentives required to tolerate these costs in North America are likely therefore to be considerably lower, and a sense of ownership and control of uses of those animals may not be required.

Another major consideration is the enforcement context in the United States and Canada compared with many other places. While there is little systematic analysis to support broad conclusions, it seems logically clear that government ownership and control of wildlife will rely for its effectiveness on strong state capacity for enforcement of restrictions on use. At least where wildlife is both valuable and accessible, enforcement capacity will be critical if *de jure* government property rights are not to result in de facto open access.[19] The resources for enforcement of wildlife laws available in Canada and the United States are orders of magnitude more abundant than in many of the world's most wildlife-rich countries. Resources for enforcement against illegal hunting of wildlife in most of the developing world are vastly inadequate. Some protected areas have few (sometimes no) vehicles, few and poorly trained staff, inadequate communications equipment, and vast areas with impassable roads. Government control of wildlife management is a pipe dream under these circumstances, and the effective arbiters of the fate of wildlife will be the communities who live close to them.

## What Is the Role of Hunting?

A second marked feature of the Model is the role of hunting in its motivation, its success, and its maintenance. Others have clearly documented the role of hunters and hunting organizations in establishing the architecture of the Model to drive species recoveries and habitat conservation,[20] as well as the ongoing importance of hunting and fishing in generating its revenue base.

The role that hunting plays in relation to conservation varies very strikingly across the globe. In Australia there is little or no formal recognition in law or policy that hunting poses anything other than a threat to conservation.[21] In India, virtually all hunting is prohibited.[22] In South Africa, it is recognized as a fundamental component of conservation strategy through providing incentives for wildlife as a land use.[23] In Tanzania, hunting is viewed as inter alia an important way to generate revenue for management of some protected areas.[24] In parts of Europe, the conservation role of hunting is related more to management of overabundant or otherwise problematic species than providing incentives and revenue for conservation.[25] These fundamental differences concerning the role of hunting reflect different historical, cultural, ecological, and policy settings, and simple prescriptions advocating "what works" in one setting to be applied in another are overly simplistic.

A key aspect of the significance of hunting in the Model, under the Public Trust Doctrine, is that anyone can benefit; the government must provide all citizens with hunting opportunities. (Of course, hunting is not the only way citizens benefit from wildlife.) This is important not only because it is based on respect for democratic principles and benefit-sharing, but because it provides a powerful motivation for conservation and recovery of game populations and enables a flow of benefits that can mobilize an enormous constituency to participate in and contribute to conservation. The history of the Model is powerful testimony to the conservation impacts this can have. Likewise, the desire to hunt or harvest wildlife populations was the major motivator for government-led efforts to restore and sustainably manage wildlife populations under the Soviet Union.[26] The positive role that recreational hunting can play in conservation (as well as its limitations) has been well evidenced elsewhere.[27]

Drawing conclusions about the success and efficacy of hunting versus nonhunting models of wildlife management is a fraught exercise, both because of the multiple factors that vary across contexts and because of the entrenched ideological positions surrounding each approach. Pack and colleagues sought to compare the outcomes of two wildlife management models based on hunting (North America, southern Africa) with two nonhunting models (Kenya, India) models of wildlife management.[28] The authors assessed the status of selected species in each example and found that the "hunting" models performed best in terms of wildlife conservation: they had more species that were stable or increasing and fewer that were declining, although these differences were not significant. Ranking all the models across a number of broader economic, social, and ecological indices, the two hunting models performed best and the North American model best overall.

A crucial struggle facing nonhunting models of wildlife management is how to induce people to care for and value wildlife, particularly outside of formal protected areas. A key dynamic driving wildlife declines is that wildlife itself or its habitat is removed or degraded in order for human economic or other activities to take place. Habitat is converted to land for crops, livestock, housing, or infrastructure; predators are persecuted in order to protect livestock or people; herbivores are killed to reduce competition with stock.[29] Thus, a critical element in a broad wildlife strategy must necessarily be to generate incentives for people to maintain wildlife and its habitat. Incentives can be financial or nonfinancial and can involve consumptive or nonconsumptive use. There are for example some excellent examples of nature-based tourism, the most common nonconsumptive approach, supporting wildlife conservation and improved local livelihoods.[30] But it is suitable only in a limited range of areas; in dense tropical forests, for example, tourism based on viewing charismatic wildlife is typically impossible. This is a difficult and central problem for conservation globally, and one of ever-increasing importance as population and consumption pressures increase demands for land in many biodiversity-rich areas.

## Can Wildlife Be Traded?

The third feature of the Model discussed here is its prohibition on commercial trade in products from dead wildlife—that is, the sale of meat, skins, or other products from hunted animals (with limited exceptions). Given that it was commercial hunting for the market that so devastated North American wildlife populations, this caution is understandable.

Caution about commercial use of wildlife is widely reflected in conservation approaches around the world. Commercial use (in the sense of hunting for trade) is frequently subject to much greater restriction at national or subnational level than subsistence or recreational use. It typically requires much more stringent licensing and monitoring. For example, bush meat trade is banned in many tropical countries, whereas provision is sometimes made for hunting for subsistence use.[31] In Australia, Indigenous people have rights to use wildlife for subsistence purposes, but not for trade.[32]

There are two concerns related to commercial as distinct from subsistence or recreational use. One is that commercial operations can mobilize social and economic pressures for larger, rather than smaller, harvests to meet market demand, potentially threatening unsustainability. Good regulation including effective enforcement is perhaps a complete answer to this problem. The other key objection is the potential for legal trade avenues to inadvertently facilitate or promote illegal trade in a number of ways. First, illegally sourced commodities can be "laundered" into legitimate trade chains to facilitate their sale, whereas without the legal trade routes the commercial opportunities for illegally sourced commodities are much more limited. This argument is relied on very extensively in debates around wildlife trade, particularly when proposals for international trade in high value commodities such as elephant ivory and rhino horn are floated. The conditions under which it holds true in practice remain opaque. Second, existence of a legal trade makes enforcement more complex: it must be determined not only whether a commodity is in possession or being traded, but whether it is of legal or illegal origin.

Despite these inherent difficulties, it is clear that commercial trade in wildlife parts does not inevitably lead to negative outcomes for the target species. In Australia, several kangaroo species are commercially harvested at the scale of up to tens of millions per year, without causing significant declines.[33] Illegal trade is insignificant. In Europe, sausages or other products from legally hunted wild boar, feral beavers, and bears are legally marketed in a number of countries. In Germany, alone, around 600,000 wild boars are hunted annually, with no detrimental population impacts (indeed, boars continue to increase).[34]

Further, commercial trade can be not only neutral but positively contribute to conservation through a number of means: increasing the value of wildlife to landowners, generating revenue for conservation, and decreasing illegal trade by providing the market with a preferred, legal product. In Namibia and South Africa, wildlife populations on private and communal land are harvested ("game cropped") for meat and skins alongside recreational hunting and tourism operations, with revenues forming part of the economic rationale for maintaining wildlife-based land uses.[35] Live trade (including of rhinos) takes place between wildlife ranches and sometimes between protected areas and ranches, as with the sale of white rhinos from Kruger National Park in South Africa to private ranches. Commercial trade in vicuña fiber (albeit live-shorn) contributes to Indigenous and local communities by reducing stock levels and reducing poaching in Andean countries. Many crocodilians globally, including American alligator in the southern United States, are harvested for commercial trade at very high levels, and where well-managed, this trade has been a positive contributor to their conservation.[36]

The underlying question of whether, and when, the existence of a legal trade makes an illegal trade more likely or jeopardizes sustainability remains contested and unresolved.[37] In many of the cases where trade currently contributes positively to species conservation, trade was restarted cautiously only after a period of strict protection to enable recovery of wildlife populations, and, in all cases, trade requires a management context that can effectively detect and respond to overharvest or illegal activity.

## Conclusion

The North American Model underpins one of the world's most significant and powerful conservation success stories. It is a unique wildlife management approach with no clear analogue globally: it has the unusual features of strong government ownership and control coupled with providing access to all and very broad rights to hunt with virtually no allowance for selling the products. In contrast to many other systems, it has mobilized incentives for conservation very effectively, in a way that does not privilege elites, that engages large sections of the populace, and that builds in a long-term funding stream. It has, of course, many other salient critical features, such as the particularly strong science base that underpins management, which have not been mentioned at all here.

While the Model faces an uncertain future and is the subject of much current debate, experiences in the United States and Canada offer a wealth of insight and lessons for conservation efforts globally. Curiously, the Model remains very poorly understood or recognized outside of these countries. As debate rages globally about hunting, sustainable use, and wildlife trade both legal and illegal, it is extraordinarily important that lessons from this experience are more widely understood.

*NOTES*


1. J. F. Organ, V. Geist, S. P. Mahoney, S. Williams, P. R. Krausman, G. R. Batcheller, T. A. Decker, R. Carmichael, P. Nanjappa, R. Regan, R. A. Medellin, R. Cantu, R. E. McCabe, S. Craven, G. M. Vecellio, and D. J. Decker, "The North American Model of Wildlife Conservation," *Wildlife Society Technical Review* 12, no. 4 (2012): 25.
2. J. R. Heffelfinger, V. Geist, and W. Wishart, "The Role of Hunting in North American Wildlife Conservation," *International Journal of Environmental Studies* 70, no. 3 (2013): 399–413; P. R. Krausman and V. C. Bleich, "Conservation and Management of Ungulates in North America," *International Journal of Environmental Studies* 70, no. 3 (2013): 372–382; D. G. Hewitt, "Hunters and Conservation and Management of White-Tailed Deer (*Odocoileus virginianus*)," *International Journal of Environmental Studies* 72, no. 5 (2015): 839–849; T. W. Hughes and K. Lee, "The Role of Recreational Hunting in the Recovery and Conservation of the Wild Turkey (*Meleagris gallopavo* spp.)," *International Journal of Environmental Studies* 72, no. 5 (2015): 797–809; K. Hurley, C. Brewer, and G. N. Thornton, "The Role of Hunters in Conservation, Restoration, and Management of North American Wild Sheep," *International Journal of Environmental Studies* 72, no. 5 (2015): 784–796.
3. Namibian Association of Community Based Natural Resource Management Support Organisations, *The State of Community Conservation in Namibia—A Review of Communal Conservancies, Community Forests and Other CBNRM Initiatives (2014/15 Annual Report)* (Windhoek, Namibia: NACSO, 2015); W. A. Taylor, P. A. Lindsey, and H. Davies-Mostert, *An Assessment of the Economic, Social and Conservation Value of the Wildlife Ranching Industry and Its Potential to Support the Green Economy in South Africa*, Green Economy Research Report (Johannesburg, Republic of South Africa: The Endangered Wildlife Trust, 2015).
4. S. Deinet, C. Leronymidou, L. McRae, I. J. Burfield, R. P. Foppen, B. Collen, and M. Böhm, *Wildlife Comeback in Europe: The Recovery of Selected Mammal and Bird Species*, final report to Rewilding Europe by ZSL, BirdLife International, and the European Bird Census Council (London: ZSL, 2013).
5. L. Baskin, "Hunting of Game Mammals in the Soviet Union," in *Conservation of Biological Resources*, ed. E. J. Milner-Gulland and R. Mace (Oxford, UK: Blackwell Science, 1998), 331–345.
6. Y. Fukuda, G. Webb, C. Manolis, R. Delaney, M. Letnic, G. Lindner, and P. Whitehead, "Recovery of Saltwater Crocodiles Following Unregulated Hunting in Tidal Rivers of the Northern Territory, Australia," *Journal of Wildlife Management* 75, no. 6 (2011): 1253–1266.
7. H. Pande and S. Arora, *India's Fifth National Report to the Convention on Biological Diversity, 2014* (New Delhi: Ministry of Environment and Forests, Government of India, 2014).
8. WWF, *Living Planet Report 2016, Risk and Resilience in a New Era* (Gland, Switzerland: WWF International, 2016).
9. S. P. Mahoney and J. J. Jackson, "Enshrining Hunting as a Foundation for Conservation—The North American Model," *International Journal of Environmental Studies* 70, no. 3 (2013): 448–459.
10. P. Shyamsundar, E. Araral, and S. Weeraratne, *Devolution of Resource Rights, Poverty, and Natural Resource Management: A Review*, Environmental Economic Series (Washington, DC: World Bank, 2005).
11. M. Cirelli, *Legal Trends in Wildlife Management*, FAO Legislative Study (Rome, Italy: Food and Agriculture Association of the United Nations, Legal Office, 2002); R. Cooney, *Commercial and Sustainable Use of Wildlife: Suggestions to Improve Conservation, Land Management and Rural Economies* (Canberra: Australian Government Rural Industries Research and Development Corporation, 2008).
12. S. Pack, R. Golden, and A. Walker, *Comparison of National Wildlife Management Strategies: What Works Where, and Why?* (Washington, DC: Heinz Centers for Science, Economics and Environment, 2013).
13. L.C. Weaver, E. Hamunyela, R. Diggle, G. Mantongo, and T. Pietersen, *The Catalytic Role and Contributions of Sustainable Wildlife use to the Namibia Community Based Natural Resource Management Programme*, CITES and CBNRM: Proceedings of an international symposium on "The Relevance of CBNRM to the Conservation and Sustainable Use of CITES-Listed Species in Exporting Countries" (Gland, Switzerland: IUCN and IIED, 2011), 59–70.
14. Cirelli, *Legal Trends*.
15. C. C. Gibson and S. A. Marks, "Transforming Rural Hunters into Conservationists: An Assessment of Community-Based Wildlife Management Programs in Africa," *World Development* 23, no. 6 (1995): 941–957.
16. Weaver et al., *The Catalytic Role*.
17. R. J. Smith, R. D. J. Muir, M. J. Walpole, A. Balmford, and N. Leader-Williams, "Governance and the Loss of Biodiversity," *Nature* 426 (2003): 263–279.
18. M. Barua, A. Bhagwat, and S. Jadhav, "The Hidden Treasures of Human-Wildlife Conflict: Health Impacts, Opportunity and Transaction Costs," *Biological Conservation* 157 (2013): 309–316.

19. N. Leader-Williams and S. Albon, "Allocation of Resources for Conservation," *Nature* 336, (1988): 503–604.
20. Heffelfinger, Geist, and Wishart, "The Role of Hunting"; Krausman and Bleich, "Conservation and Management"; Mahoney and Jackson, "Enshrining Hunting"; C. Semcer and J. Posewitz, "The Wilderness Hunter: 400 Years of Evolution," *International Journal of Environmental Studies* 70, no. 3 (2013): 438–447; M. G. Anderson and P. I. Padding, "The North American Approach to Waterfowl Management: Synergy of Hunting and Habitat Conservation," *International Journal of Environmental Studies* 72, no. 5 (2015): 810–829; E. B. Arnett and R. Southwick, "Economic and Social Benefits of Hunting in North America," *International Journal of Environmental Studies* 72, no. 5 (2015): 734–745; P. Krausman and S. P. Mahoney, "How the Boone and Crockett Club (B&C) Shaped North American Conservation," *International Journal of Environmental Studies* 72, no. 5 (2015): 746–755.
21. Cooney, *Commercial and Sustainable Use*.
22. M. Misra, "Evolution, Impact, and Effectiveness of Domestic Wildlife Trade Bans in India," in *The Trade in Wildlife: Regulations for Conservation*, ed. S. Oldfield (London: Earthscan, 2003), 78–85.
23. Department of Environmental Affairs, *Biodiversity Economic Strategy (BES) for the Department of Environmental Affairs*, No. 39268 (Pretoria, Republic of South Africa: R.o.S.A. Department of Environmental Affairs, 2015); Department of Environmental Affairs and Tourism, *South Africa's National Biodiversity Strategy and Action Plan* (Pretoria, South Africa: Department of Environmental Affairs and Tourism, 2005).
24. International Union for Conservation of Nature, *Informing Decisions on Trophy Hunting* (Gland, Switzerland: International Union for Conservation of Nature, 2016).
25. G. Cedurland, J. Bergqvist, P. Kjellander, and P. Duncan, "Managing Roe Deer and Their Impact on the Environment: Maximizing the Net Benefits to Society," in *The European Roe Deer: The Biology of Success*, ed. R. Anderson, P. Duncan, and J. D. C. Linell (Oslo, Norway: Scandinavian University Press, 1998), 337–372; H. Geiser, H. U. Reyer, and P. Krausman, "Efficacy of Hunting, Feeding, and Fencing to Reduce Crop Damage by Wild Boars," *Journal of Wildlife Management* 68, no. 4 (2004): 939–946; J. M. Milner, C. Bonenfant, A. Mysterud, J. Gaillard, S. Csányi, and N. C. Stenseth, "Temporal and Spatial Development of Red Deer Harvesting in Europe: Biological and Cultural Factors," *Journal of Applied Ecology* 43, no. 4 (2006): 721–734.
26. Baskin, "Hunting of Game Mammals."
27. B. Dickson, J. Hutton, and W. Adams, eds., *Recreational Hunting, Conservation and Rural Livelihoods: Science and Practice* (Oxford, UK: Wiley Blackwell, 2003); IUCN, *Informing Decisions*.
28. Pack, Golden, and Walker, *Comparison of National Wildlife Management Strategies*.
29. G. Mace, H. Masundire, and J. Braille, "Biodiversity," in *Ecosystems and Human Well-Being: Current State and Trends: Findings of the Condition and Trends Working Group*, ed. R. Hassan, R. Scholes, and N. Ash (Washington, DC: Island Press, 2005), 77–122.
30. O. Krüger, "The Role of Ecotourism Is Conservation: Panacea or Pandora's Box?" *Biodiversity and Conservation* 14, no. 3 (2005): 579–600; A. Stronza and F. Pêgas, "Ecotourism and Conservation: Two Cases from Brazil and Peru," *Human Dimensions of Wildlife* 13, no. 4 (2008): 263–279; C. A. Hunt, W. H. Durham, L. Driscoll, and M. Honey, "Can Ecotourism Deliver Real Economic, Social, and Environmental Benefits? A Study of the Osa Peninsula, Costa Rica," *Journal of Sustainable Tourism* 23, no. 3 (2015): 339–357.
31. Collaborative Partnership on Sustainable Wildlife Management, *CPW Factsheet: Sustainable Wildlife Management and Wild Meat* (Rome, Italy: UN Food and Agriculture Organization, 2014).
32. R. Cooney and M. Edwards, *Indigenous Wildlife Enterprise Development: The Regulation and Policy Context and Challenges*, Report to North Australian Indigenous Land and Sea Management Alliance (Darwin, Australia: North Australian Indigenous Land and Sea Management Alliance [NAILSMA], 2008).
33. R. Cooney, M. A. Archer, A. Baumber, P. Ampt, G. Wilson, J. Smitts, and G. Webb, "THINNK Again: Getting the Facts Straight on Kangaroo Harvesting and Conservation," in *Science under Siege: Zoology under Threat*, ed. P. Banks, D. Lunney, and C. Dickman (Mosman, Australia: Royal Zoological Society of New South Wales, 2012), 150–160.
34. O. Keuling, E. Strauss, and U. Siebert, "Regulating Wild Boar Populations Is 'Somebody Else's Problem!'—Human Dimension in Wild Boar Management," *Science of the Total Environment*, no. 554 (2016): 311–319.
35. Weaver et el., *The Catalytic Role*, 59–70.
36. J. M. Hutton and G. Webb, "Crocodiles: Legal Trade Snaps Back," in *The Trade in Wildlife: Regulation for Conservation*, ed. S. Oldfield (London: Earthscan, 2003), 108–120.
37. E. H. Bulte, R. Damania, and G. C. Van Kooten, "The Effects of One-Off Ivory Sales on Elephant Mortality," *Journal of Wildlife Management* 71, no. 2 (2007): 613–618; J. Phelps, L. Carrasco, and E. L. Webb, "A Framework for Assessing Supply-Side Wildlife Conservation," *Conservation Biology* 28, no. 1 (2013): 244–257.

# 13 The Model in Transition

## *From Proactive Leadership to Reactive Conservation*

Shane P. Mahoney

*As the previous writings have illuminated, the North American Model emerged through a process of organic change. No group or individual ever wrote a prescription for this conservation approach, nor did anyone forecast the many dimensions that would ultimately become integral to it. Certainly, no judicial process ordained it, nor did any political manifesto proclaim it. Rather, it was compelled to prominence by the same vague processes that seem to guide all social progress or disintegration, including the erosion of existing values and perspectives by tidal forces that seemingly have little in common but find partnership in drifting, meandering ways. Economic, political, and cultural dimensions of societies may, at times, appear to operate independently but in truth they are in constant interaction and are ecologically entwined. The change they experience is the change they, in fact, are making, even as they offer resistance to it. This chapter will succinctly review the Model's past, present, and future in a changing society.*

### Introduction

Like all such processes, the Model's emergence and its developing principles were initially dismissed as fringe activism. The growing social discourse concerning wildlife's future that gave rise to the Model would not, however, be quieted. It would eventually be embraced by the journalistic mainstream and then enter the corridors of political power. The gradual extension of its social frontier was likely imperceptible to most observers, perhaps even to its protagonists. Nevertheless, change was underway, and viewed in hindsight, it illustrated the capacity of awareness campaigning to disrupt social norms, nourish incipient movements, and lead to directional change in public opinion. The Model's emergence was undoubtedly facilitated by one of the earliest educational efforts in conservation history, though few perhaps recognized this at the time or even today. Nevertheless, it was the cumulative agitation and efforts on the part of hunting clubs, wilderness proponents, magazine editors, and newspaper columnists, as well as the salon discourse of social elites, public lecturers, and book writers of the day that brought the desperate fate of the continent's wildlife to the public conscience. Perhaps no one could say when the tipping point occurred, nor identify when the public's awareness finally came to matter.

Indeed, it would be difficult to say when exactly the North American Model emerged or was finally constructed. One thing is certain, however; it did not arise fully formed, nor could anyone have imagined the many components that today are recognizable and considered essential to it. Even obvious candidates of seeming necessity, such as provincial and state wildlife agencies, were not forecast or predetermined. Rather, the Model emerged furtively at first, contesting prevailing value propositions that saw wildlife as limitless and those unrestricted economic derivatives that arose from it as a right of citizenship. Its maturation to an integrated and near-continental complex of laws, policies, and institutions that placed restrained use for primarily noneconomic purpose as its basis was indeed a remarkable innovation for that time, or for any other. The social forces that launched this approach were primarily concerned with wildlife depletion and loss, and it was the reality of wildlife decline that initially motivated them and brought relevance to defining a new relationship between progress, citizenship, and natural resource use.

## Proactive Leadership

The vanguards of change were not just content with halting wildlife declines. What they sought was recovery to abundance and an enduring commitment to wise use. This was a bold, startling departure. It did not struggle to find workings within the existing framework of wildlife valuation and use (reactive agency) but offered a startling new definition of what was possible (proactive visioning). The early leaders were equally energetic in promoting landscape preservation and safeguarding prized natural areas from destruction and, sometimes, from any form of use at all, except the harvest of beauty by the awestruck traveler. Both wilderness aesthetics and wildlife pursuits required wildlife recovery, of course, but it was the former that helped open a wider social channel of awareness that included women as influential proponents for change. Thus did the conservation community expand in formation, conscripted by the new vision that opened doors while changing minds. Few women at that time hunted but recreational pursuits such as hiking, bird-watching, and boating were popular activities. Today, the number of women participating in hunting is increasing.[1] Perhaps their growing influence may help reunite the founding philosophies of North American conservation, which diverged, sometimes to the point of antagonism, over the last hundred years.

This is not to suggest that proponents for inclusive conservation disappeared entirely in the early aftermath of the movement's emergence. On the contrary, there were influential proponents of this wider view who appeared on the scene a generation after the Model's founding in the late nineteenth and early twentieth centuries. Echoes of their inheritance reverberate still, though an increasingly polarized politic in conservation now weakens the middle ground they staked and occupied. Certainly, Aldo Leopold remains one of the Model's most iconic personas, and his enduring influence derives both from his philosophical writings and inclusive view of man, land and wildlife and from his detailed prescriptions for managing wildlife as a sustainable natural resource.[2] A hunter and nature lover, Leopold has maintained credibility as a representative of both dimensions, though this integrated dichotomy seems increasingly difficult to attain in modern social perceptions. Like landscape preservation and hunting, a love of nature and hunting have been increasingly viewed as separate paths, though it is clear that maintaining such perceptions can only weaken conservation and the Model itself. The North American Model today is being weakened by such divisions, yet at one time, indeed for a long time, such divisions did not exist. As hunting participation falters in North America and as distance from nature increases for larger numbers of people, shared necessity may encourage a return to the stronger coalitions of the past. In a world where too few people have concern for conservation, uniting those who do must become a priority.

But more than this needs to happen if wildlife is to thrive and traditional human engagements with it are to survive in North America. What is certain is that to function as a conservation mechanism, the Model must adapt to changing social circumstances and to the new ecological realities brought on by climate change and the expanding human footprint that seeks space and pursues natural resource extraction at escalating rates.[3] And adaptation is not the same as reaction, a distinction that seems to elude much of the current thinking on the Model and North American conservation, in general. What is most striking in reviewing conservation's rise and progress in North America is just how strategic its early leaders were in their approach to crisis and challenge. Faced with immense difficulties, they did not falter to reactionary approaches but sought long-term mechanisms that would align with human incentives such as citizenship responsibility, personal sustenance, enjoyment, and a love of nature. Whether this was by design, happy accident, or a combination of both, who can say? Regardless, the principles they identified and the vision they promoted were designed to long outlive the tenure of their own lives and circumstances. Indeed, their emphasis on future generations was one of the most prevalent and inspiring elements of their thought process. Perhaps it was this, more than any other principle, that guided their conservation worldview and helps explain why their efforts gave rise to a conservation Model that would last more than a century. If we wish to define a new or revised approach to conservation in North America, we should reflect more on what these early leaders did and seek guidance from their approach.

Consider, in this regard, the early emphasis on hunting and the establishment of protected areas. At the time, excessive killing was the major cause of wildlife decline, and bringing this to a halt was the most critical issue before them. A more reactionary approach might have reasonably called for the elimination of all harvesting, but the early leaders did not, in fact, seek to end wildlife harvest; rather, they made it central to the recovery and maintenance of populations and species. With regard to protected areas, the early efforts to safeguard special places such as national parks took place far in advance of any national wilderness crisis. Indeed, the earliest efforts occurred when westward expansion of European settler populations was still fairly modest and when vast expanses of the continent remained in their natural state. Both of these examples point to the long-term, strategic vision developed by the leading conservation activists of the day. We could also note the very early call for empirical observation and scientific specialist advice to guide conservation practice as one more example of the longer-term vision of the Model's early leaders.

Despite such strategic visioning, the Model is not without its weaknesses,[4] nor should it be assumed, that it is the best Model we can devise. Neither the Model as it exists, however, nor any that might replace it, should be the primary objective for those focused on conservation success; and seeking guidance from early leaders should not be pursued simply to defend, diminish, or make a better version of the Model they developed. The objective of debate must be to achieve the very best possible outcomes for wildlife in North America and to safeguard and expand the incentivized public who will enjoy, defend, and care for experiences in nature. In this regard, both defenders and critics of the Model are faced with a similar challenge: how to establish a vision for wild places and wild creatures in North America that also helps create a conservation-supportive society. Neither challenging a principle of the Model nor defending it will necessarily make wildlife's future more secure or safeguard the successes of past conservation efforts. What we need is a revitalized and reconstituted conservation movement that is uncompromising in its values but flexible regarding the pathways to success. What model society chooses to achieve this is a secondary issue. In this instance, the "how" is important, but the "what" and "why" are critical.

## Reactive Conservation

Yet, it is hard to escape the impression that, increasingly, conservation debate in North America is mostly reactive, focused on issues relevant to the movement as it currently exists. Less proactive in its visioning overall, it is also less inclined to seek a new order of things and more politicized in its demeanor than ever before. This is most regrettable, as North America retains so many advantages that predispose it to achieving even greater things for wildlife and conservation. Democracy, rule of law, wealth, education, abundant wildlife, and other natural resources, mature institutions, highly evolved legislation and policies, expansive landscapes, and relatively few people are all advantages the United States and Canada share. What we lack is a clear, bold vision that describes what may be wrong and what needs to be corrected and outlines the path forward to a better world for wildlife and our human needs. Neither particulate criticisms of the status quo nor ideological defenses of it provide helpful guidance. Certainly, neither should be confused with vision.

It was exactly this kind of bold visioning that the leadership in Canada and the United States launched over one hundred year ago. Much has changed since that time, and thoughtful reevaluation of our conservation Model is required. So, too, is a rationale for modern times. We now require an explanation of conservation urgency and responsibility in terms meaningful to the world we live in, not the world we were born to. Modern North America requires a modern conservation idiom, one that encourages inclusivity, energizes perseverance, and establishes rituals and metrics for success. Without commitment to fundamental change, we must accept that current trends in conservation will continue and that the North American Model will inevitably weaken as its foundational principles prove outdated to both nature's requirements and society's tolerances and values.

Some might say this is a good thing, but what alternative do we have available to us? What is our next conservation play? Who has this blueprint, and how will it be moved to influence? Who are the architects of power, the mobilizers of knowledge? Who will be the heroes every movement needs? We might remind those seeking change but offering no detailed plan for reconstruction that movements, like organizations, should adapt when they are strong, not react when they are weakened. And to those who resist change on principle, we might note that all retreats to fortress are, eventually, doomed. All of us might reflect that change, both in nature and society, follows a pattern of punctuated equilibrium where, eventually, precipitous change overtakes and resets the conditions for longer periods of relative stasis.[5] Very likely, North American conservation stands at this precipice now. Our decision must be whether to guide such change, simply demand that it occurs, or deny that it is needed. In such a circumstance, what real choice does any of us have?

## Conclusion

Many countries today are identified so closely with the wild creatures and landscapes they harbor that losing them ought to be seen as a national crisis. Canada and the United States are among these. Yet, both are nation states that have embraced continuous economic growth as a foundational policy, regardless of the rhetorical fallacies fashioned by political elites. Not surprisingly, both countries face substantial conservation challenges. Any reasonable forecast based upon their current worldview would suggest such challenges will increase. As human populations increase and resource extraction escalates, no amount of reactionary conservation can be expected to do more than slow the inevitable decline in nature's diversity and our own standards of life and experience that will accompany it.

Wild nature cannot speak for itself. We must. The greatest question facing North American society today is whether we will. If any North American model is to succeed, it will make a visionary stand that repositions the future of wild others and the places they require as a responsibility of citizenship. In this context, the debates over the existing Model seem destined to wither to oblivion, much like so many wild species and wild places already have.

*NOTES*

1. K. A. Schmitt, "More Women Give Hunting a Shot," *National Geographic*, November 4, 2013, https://news

.nationalgeographic.com/news/2013/11/131103-women-hunters-local-meat-food-outdoor-sports/.
2. A. Leopold, *A Sand County Almanac* (New York: Oxford University Press, 1949); A. Leopold, *Game Management* (New York: Charles Scribner's Sons, 1933).
3. M. Grooten and R. E. A. Almond, eds., *WWF, Living Planet Report, 2018: Aiming Higher* (Gland, Switzerland: WWF, 2018).
4. M. J. Peterson and M. P. Nelson, "Why the North American Model of Wildlife Conservation Is Problematic for Modern Wildlife Management," *Human Dimensions of Wildlife* 22 (2016): 43–54; J. A. Vucetich, P. C. Paquet, and J. K. Bump, "An Inadequate Construct? North American Model: What's Flawed, What's Missing, What's Needed," *Wildlife Professional* 5 (Summer 2013): 58–60; K. A. Artelle, J. D. Reynolds, A. Treves, J. C. Walsh, P. C. Paquet, and C. T. Darimont, "Hallmarks Missing from North American Wildlife Management," *Science Advances* 4, no. 3 (2018): eaao0167.
5. S. J. Gould, *Punctuated Equilibrium* (Cambridge, MA: The Belknap Press of Harvard University Press, 2007).

# Index